Thorsten Vehoff

Simulations of Charge Transport in Organic Compounds

Thorsten Vehoff

Simulations of Charge Transport in Organic Compounds

Südwestdeutscher Verlag für Hochschulschriften

Imprint
Any brand names and product names mentioned in this book are subject to trademark, brand or patent protection and are trademarks or registered trademarks of their respective holders. The use of brand names, product names, common names, trade names, product descriptions etc. even without a particular marking in this work is in no way to be construed to mean that such names may be regarded as unrestricted in respect of trademark and brand protection legislation and could thus be used by anyone.

Publisher:
Südwestdeutscher Verlag für Hochschulschriften
is a trademark of
Dodo Books Indian Ocean Ltd., member of the OmniScriptum S.R.L Publishing group
str. A.Russo 15, of. 61, Chisinau-2068, Republic of Moldova Europe
Printed at: see last page
ISBN: 978-3-8381-2081-2

Zugl. / Approved by: Mainz, Johannes Gutenberg-Universität Mainz, Diss., 2010

Copyright © Thorsten Vehoff
Copyright © 2010 Dodo Books Indian Ocean Ltd., member of the OmniScriptum S.R.L Publishing group

Abstract

To aid the design of organic semiconductors, we study the charge transport properties of organic liquid crystals, i.e. hexabenzocoronene and carbazole macrocycle, and single crystals, i.e. rubrene, indolocarbazole and benzothiophene derivatives (BTBT, BBBT). The aim is to find structure-property relationships linking the chemical structure as well as the morphology with the bulk charge carrier mobility of the compounds. To this end, molecular dynamics (MD) simulations are performed yielding realistic equilibrated morphologies. Partial charges and molecular orbitals are calculated based on single molecules in vacuum using quantum chemical methods. The molecular orbitals are then mapped onto the molecular positions and orientations, which allows calculation of the transfer integrals between nearest neighbors using the molecular orbital overlap method. Thus we obtain realistic transfer integral distributions and their autocorrelations.

In case of organic crystals the differences between two descriptions of charge transport, namely semi-classical dynamics (SCD) in the small polaron limit and kinetic Monte Carlo (KMC) based on Marcus rates, are studied. The liquid crystals are investigated solely in the hopping limit. To simulate the charge dynamics using KMC, the centers of mass of the molecules are mapped onto lattice sites and the transfer integrals are used to compute the hopping rates. In the small polaron limit, where the electronic wave function is spread over a limited number of neighboring molecules, the Schrödinger equation is solved numerically using a semi-classical approach. The results are compared for the different compounds and methods and, where available, with experimental data. The carbazole macrocycles form columnar structures arranged on a hexagonal lattice with side chains facing inwards, so columns can closely approach each other allowing inter-columnar and thus three-dimensional transport. When taking only intra-columnar transport into account, the mobility is orders of magnitude lower than in the three-dimensional case. BTBT is a promising material for solution-processed organic field-effect transistors. We are able to show that, on the time-scales of charge transport, static disorder due to slow side chain motions is the main factor determining the mobility. The resulting broad transfer integral distributions modify the connectivity of the system but sufficiently many fast percolation paths remain for the charges. Rubrene, indolocarbazole and BBBT are examples of crystals without significant static disorder. The high mobility of rubrene is explained by two main features: first, the shifted cofacial alignment of its molecules, and second, the high center of mass vibrational frequency.

In comparsion to SCD, only KMC based on Marcus rates is capable of describing neighbors with low coupling and of taking static disorder into account three-dimensionally. Thus it is the method of choice for crystalline systems dominated by static disorder. However, it is inappropriate for the case of strong coupling and underestimates the mobility of well-ordered crystals. SCD, despite its one-dimensionality, is valuable for crystals with strong coupling and little disorder. It also allows correct treatment of dynamical effects, such as intermolecular vibrations of the molecules. Rate equations are incapable of this, because simulations are performed on static snapshots.

We have thus shown strengths and weaknesses of two state of the art models used to study charge transport in organic compounds, partially developed a program to compute and visualize transfer integral distributions and other charge transport properties, and found structure-mobility relations for several promising organic semiconductors.

Contents

1. **Introduction**   1
2. **Organic Electronics**   5
   - 2.1. Organic Electronic Devices 5
     - 2.1.1. Organic Solar Cells 5
     - 2.1.2. Organic Light Emitting Diodes 8
     - 2.1.3. Organic Field Effect Transistors 9
   - 2.2. Measurements of Charge Mobility 11
     - 2.2.1. Time of Flight 11
     - 2.2.2. Pulse Radiolysis - Time Resolved Microwave Conductivity (PR-TRMC)   12
     - 2.2.3. Field effect transistor and diode measurements 13
3. **Computer simulations**   15
   - 3.1. Ab initio quantum chemistry 15
   - 3.2. Density Functional Theory (DFT) 16
   - 3.3. Molecular dynamics simulations 16
     - 3.3.1. Force fields 16
     - 3.3.2. Calculation of force field parameters 17
     - 3.3.3. Integration schemes 21
     - 3.3.4. Thermostats 22
     - 3.3.5. Barostats 26
   - 3.4. Kinetic Monte Carlo (KMC) 27
     - 3.4.1. Infrequent-event systems 27
     - 3.4.2. The KMC procedure 29
     - 3.4.3. Determination of the rate constants 29
4. **Models for charge transport**   31
   - 4.1. Band theory 31
   - 4.2. Polaron models 33
     - 4.2.1. Holstein models 33
     - 4.2.2. Holstein-Peierls models 35
   - 4.3. Diffusion Limited by Thermal Disorder 36
     - 4.3.1. Semi-classical Dynamics (SCD) 36
   - 4.4. Disorder models 41
     - 4.4.1. Miller-Abrahams rates 41
     - 4.4.2. Marcus rates 42
     - 4.4.3. Gaussian disorder model (GDM) 45
     - 4.4.4. Kinetic Monte Carlo with Marcus rates 46

Contents

- 4.5. Charge transport parameters ... 49
  - 4.5.1. Site energy ... 49
  - 4.5.2. Reorganization energy ... 50
  - 4.5.3. Transfer integrals ... 50
  - 4.5.4. Change in free energy ... 52

5. Carbazole macrocycle — 55
- 5.1. Force field parameters ... 56
- 5.2. Molecular dynamics simulations ... 63
- 5.3. Charge transport parameters ... 65
- 5.4. Rate-based simulations of charge dynamics ... 67
  - 5.4.1. Orbital mapping and modeling of intra-chain transport ... 67
  - 5.4.2. Transfer integral distributions & cut-off radius ... 68
  - 5.4.3. Charge mobility ... 69
  - 5.4.4. Conclusions ... 73

6. [1]Benzothieno[3,2-b]benzothiophene (BTBT) — 75
- 6.1. Force field and MD simulations ... 76
- 6.2. Charge transport using Marcus rates ... 77
- 6.3. Charge transport using semi-classical dynamics ... 81
- 6.4. Conclusions ... 83

7. Organic Crystals — 85
- 7.1. Rubrene ... 86
  - 7.1.1. Molecular dynamics simulations and charge transport parameters ... 87
  - 7.1.2. Linking structure and transfer integral distributions ... 87
  - 7.1.3. Charge transport simulations ... 88
- 7.2. Indolocarbazole ... 89
  - 7.2.1. Molecular dynamics simulations and charge transport parameters ... 90
  - 7.2.2. Linking structure and transfer integral distributions ... 91
  - 7.2.3. Charge transport simulations ... 92
- 7.3. Benzo[1,2-b:4,5-b']bis[b]benzothiophene (BBBT) ... 94
  - 7.3.1. Molecular dynamics simulations and charge transport parameters ... 94
  - 7.3.2. Linking structure and transfer integral distributions ... 95
  - 7.3.3. Charge transport simulations ... 99
- 7.4. Charge transport using rate equations ... 99
  - 7.4.1. Cluster analysis ... 101
  - 7.4.2. Comparison of experimental and theoretical results ... 102
  - 7.4.3. Conclusions ... 103
- 7.5. Charge transport via semi-classical dynamics ... 105
- 7.6. Discussion ... 107

8. Hexabenzocoronene — 109
- 8.1. Compounds and molecular dynamics simulations ... 110
- 8.2. Charge transport parameters ... 112
- 8.3. Rate-based simulations of charge dynamics ... 113

8.4.	Conclusions	116

9. Discussion   119
- 9.1. Charge transport in columnar systems 119
- 9.2. Static disorder 120
- 9.3. Dimensionality 120
- 9.4. Ideal morphologies 121
- 9.5. SCD vs. rate-based descriptions 121
- 9.6. Future possibilities 122

A. Appendix   123
- A.1. Quantum Chemistry 123
  - A.1.1. Variational principle 123
  - A.1.2. Born-Oppenheimer approximation 124
  - A.1.3. Linear combination of atomic orbitals (LCAO) 125
  - A.1.4. Self-consistent field (SCF) 125
  - A.1.5. Slater determinant 126
  - A.1.6. Restricted Hartree Fock 127
  - A.1.7. Semi-empirical methods 129
  - A.1.8. Basis sets 130
  - A.1.9. Post Hartree Fock 132
  - A.1.10. Perturbation Theory 134
- A.2. Density Functional Theory (DFT) 135
  - A.2.1. Thomas Fermi theory 135
  - A.2.2. The Hohenberg Kohn theorem 136
  - A.2.3. Applications of DFT 137
  - A.2.4. The Kohn-Sham equations 137
  - A.2.5. Construction of exchange functionals 141
- A.3. Force field parameterization 143
  - A.3.1. [1]Benzothieno[3,2-b]benzothiophene (BTBT) 143
  - A.3.2. Indolocarbazole 144
  - A.3.3. Benzo[1,2-b:4,5-b']bis[b]benzothiophene (BBBT) 146

Acknowledgments   159

1. Introduction

In a time where the world's ever-growing energy consumption faces the reality of limited and quickly diminishing resources, it is one of the biggest challenges for science to at first reduce the current per head energy consumption and to ultimately find an alternative inexhaustible method to generate electric power. Reduction of energy consumption should be achieved by maintaining the same living standards at a lower cost, such as by reducing the cost of heating due to better insulation of buildings or by inventing more energy-efficient devices. The invention of light emitting diodes (LEDs) stands out as a low energy alternative to traditional incandescent light bulbs for household lighting. The efficiency already surpasses that of any modern energy-saving halogen light bulb. However, the reproduction of warm white light creating a comfortable living atmosphere by mixing the light of monochromatic LED sources is as of now still under development. Unfailing energy sources are plentiful in nature: the tides of the ocean, the flow of the rivers, the blow of the wind and the light of the sun. Hydropower is already in use very successfully but has almost reached its peak production without covering a significant amount of any country's energy consumption. Windmills not only severely diminish the beauty of any landscape, they are also incapable of producing a reliable amount of energy since they dependend on the everchanging strength of the wind and are efficient only in the best geographical locations. The sun, however, shines every day and its light delivers an energy of $4.3 \cdot 10^{20}$ J per hour to earth, which exceeds the total energy consumed by humanity per year, estimated to be $4.1 \cdot 10^{20}$ J [1]. At present the most efficient way to generate electrical power from sunlight is to heat a fluid and use its heat to run a conventional thermal power plant. Use of state-of-the-art inorganic solar cells is limited by their extremely high production costs mostly due to the requirement of ultra pure silicon layers. Nonetheless, solar power appears as the most promising technology to meet the challenges of the future.

Solar cells, light-emitting diodes and thin film transistors based on organic molecules show promise in solving the above difficulties. While the efficiency of organic solar cells will probably never reach that of single-crystalline silicon, the efficiency of organic devices has increased rapidly over the last years. Most importantly the low cost of synthesizing organic molecules as well as the availability of solution processing or inkjet printing as an alternative to high vacuum vapor deposition for device fabrication make organic electronic devices powerful competitors for their inorganic predecessors. Organic molecules may also be designed to absorb or emit any given wavelength, showing promise to solve the problem of building a white LED. The main challenge in improving the performance of organic solar cells to rival their inorganic counterparts is the severe lack of understanding of the underlying chemical and physical processes. Despite the ability to design countless different molecules, it is unlikely that rapid advances will be made on a trial-and-error basis. Thus it is important to gain a deeper understanding of relations between the chemical molecular structure, the corresponding morphology and finally the properties of the resulting film or device. Areas of interest are the study of self-assembly, morphology and transport properties at interfaces and in the bulk. In this work we develop structure-property relationships based on the chemical structure of specific organic molecules and their corresponding bulk morphologies. We then derive the charge transport properties of the compound without use of fitting parameters. While parameters certainly enter in quantum chemical computations and molecular dynamics force fields, the resulting transport properties are obtained without fitting any parameters to experimental data. Only the chemical structure of the molecule and the x-ray structure of the morphology are taken from experiments. Thus we use a parameter-free description of charge transport.

The starting point for every study conducted in this thesis is the chemical structure of a molecule of interest found in recent literature on organic electronics or synthesized by our experimental collaborators. The single molecule in vacuum is used to calculate the equilibrium geometry as well as the molecular orbitals by use of quantum chemical methods. The force field parameters required for molecular dynamics (MD) simulations are taken from existing force fields or computed via ab initio and density functional theory in case they are not available there. To ensure a realistic bulk morphology, x-ray data is used to generate a starting configuration. The system is then equilibrated at room temperature using molecular dynamics allowing for small volume adjustments. Parameters required for charge transport simulations such as transfer integrals, their autocorrelation functions, electrostatics etc. are calculated based on multiple snapshots of the equilibrated morphology. Transfer integral calculations for such a large number of differently aligned neighboring pairs has only recently become a feasible task by use of the molecular orbital overlap (MOO) method [2]. Most other necessary charge transport parameters are calculated using the charge transport extension of the VOTCA package (www.votca.org), partially written in the course of this thesis. Finally two different approaches to simulate charge transport are treated. The first, a three-dimensional kinetic Monte Carlo method based on Marcus rates, assumes that there is strong disorder in the system leading to the localization of a charge carrier on a specific molecule so that charge transport proceeds by a hopping process. The second, diffusion limited by thermal disorder via one-dimensional semiclassical dynamics, computes the spreading of an electronic wave function, initially localized on a few molecules, over time and is based on a model Hamiltonian taking into account disorder due to inter and intra molecular vibrations. The studies give insights not only on structure-property relations but also on shortcomings and advantages of the two different theoretical approaches.

The thesis is organized as follows. First, we will give an overview over the different organic electronic devices, which rely good charge carrier mobilities, i.e. organic solar cells, light emitting diodes and field effect transistors. We also introduce three prominent experimental techniques to measure the mobility: time of flight and pulse radiolysis time resolved microwave conductivity stand in contrast to the device-based field effect transistor measurements and give reasons why obtained mobilities may vary by orders of magnitude between the methods. In chapter three we give an overview of the theoretical basis for our simulations by giving a brief overview of the most important quantum chemical concepts. We also cover molecular dynamics simulations with an emphasis on force field parameterization and conclude the chapter with the concept of kinetic Monte Carlo. Chapter four deals with the history and recent developments in the area of charge transport theory for organic compounds. It contains the basic principles underlying band and hopping transport, but also gives specific implementation details on methods used and partially developed in this work. Chapters five to eight deal with the application of our methods to different systems. In case of the π-conjugated carbazole macrocycle we study the effect of conjugation length along the cycle on the mobility. We do so by looking at the two extreme cases of a fully conjugated cycle and one where the conjugation is broken after each monomer unit. We chose to study [1]Benzothieno[3,2-b]benzothiophene (BTBT), a solution-processable compound for organic field effect transistors, to emphasize the importance of static disorder on organic crystals introduced by slow side chain motions. Disorder and resulting percolation paths are visualized using a connectivity graph showing transfer integral values present in the system. In the chapter on organic crystals we compare four different organic crystalline bulk morphologies, neither of which is dominated by static disorder. We compare 1D, 2D and 3D transporting systems and also elaborate on the influence of different alignments between neighboring molecules. In both the chapter on organic crystals and on BTBT we link transfer integral distributions with the corresponding nearest neighbor pairs in the morphology and are able to differentiate between static and dynamic disorder. Advantages and drawbacks of semi-classical dynamics and kinetic Monte Carlo via Marcus rates are compared for all crystalline compounds. For hexabenzocoronene derivatives we continue the studies of transport in liquid crystalline columns by elaborating on the differences between a hexabenzocoronene core with four different side chains, which lead to different ordering in the morphology and hence varying transport properties. Finally, we summarize insights gained on factors influencing charge transport, conclude our comparison of the two simulation approaches and propose future work in the discussion.

2. Organic Electronics

2.1. Organic Electronic Devices

In the early 20th century all electronic devices were made of metals and conductivity was believed to exist only in well-ordered crystalline systems were the electronic wave function was periodically spread over the entire crystal and charge transport could be understood by band theories. Due to increased interest in organic materials, polymers and plastics during the second world war, electrical conductivity was also discovered in organic molecules. For polymers it was found in 1976 by oxidizing polyacetylene with bromine or chlorine by Hideki Shirakawa, Alan Mac Diarmid and Alan Heeger who received the Nobel prize in chemistry for this work in 2000 [3]. Since then research on organic electronic devices has steadily increased and they have begun to replace their inorganic counterparts for certain applications today. Inorganic devices require highly-purified silicon and vapor-deposition techniques in vacuum to be efficient. Organic materials on the other hand do not require rare and expensive elements, devices may be built by solution-processing or inkjet-printing, possess high flexibility and above all promise to be cheaper once they are at the mass production stage. However, at present they still fail to reach the performance of inorganic devices, but the great amount of research currently invested in the area is likely to change this in the near future. Even today they are capable of outperforming inorganic devices in areas where high flexibility or large area processing are required. In the following we will cover the three most important organic electronic devices: solar cells, light emitting diodes and thin film transistors. Their working principles are explained and current challenges highlighted.

2.1.1. Organic Solar Cells

Prior to treating the peculiarities of organic solar cells we shall quickly discuss their inorganic counterparts. Conventional inorganic solar cells are p-n-junctions, i.e. diodes. They are designed as depicted on the left of fig. 2.1. Usually they are made of a thin n-doped layer and a thick p-doped layer. The top layer is thin to allow as much light as possible to reach the p-n-junction, the bottom layer is thick to absorb all incoming light. At the p-n junction excess carriers will recombine, i.e. the excess electrons of the p-type material with the excess holes of the n-type material. This leads to an area of fixed charges near the p-n junction, the so-called 'space charge region'. Since there will be negative space charges in the p-doped region and positive space charges in the n-doped, this gives rise to a potential difference or voltage between the p- and n-doped regions and hence an electric field. The working principle of the solar cell is as follows: Incoming light is absorbed in the space charge region of the solar cell. The resulting electron-hole pair is immediately separated due to the electric field in the region and the fact that is weakly bound due to high screening in metals. Consequently, the electrons and holes travel toward the opposing metal contacts. Note that the electrons travel toward the contact on

2. Organic Electronics

the n-doped side, which is the opposite direction of where they would travel, if the device were functioning as a diode with externally applied voltage. The resulting current-voltage relationship is shown for an organic solar cell in fig. 2.1, but it is almost identical to that of an inorganic cell. The I-V curve of the illuminated solar cell resembles that of a diode (dark current) shifted down. Both inorganic and organic solar cells are characterized by the following properties:

- Open circuit voltage U_{oc}
  The voltage measured when the solar cell is not integrated into an electric circuit. In inorganic solar cells this corresponds to counterbalancing the internal potential of the space charge region, while in organic solar cells it cooresponds to the potential difference due to different work functions of the electrodes. The intersection between dark and illuminated curve is higher, because in organic solar cells the potential has to additionally overcome the polaron binding energy.

- Short circuit current I_{sc}
  The current measured if the solar cell is part of an electric circuit with negligible resistance. For the inorganic solar cell this is also the maximum current, for the organic solar cell the maximum may be at negative voltages if polaron pair dissociation is difficult. This may be the case for thick active layers, which reduce the internal field between the electrodes.

- Maximum power point (MPP)
  The point of operation at which the solar cell produces its maximum power output, i.e. where the product of voltage and current is at its maximum as indicated by the orange square in fig. 2.1.

- Fill factor (FF)
  The fill factor is the quotient between the maximum power and the product of open circuit voltage and short circuit current, i.e. it measures the 'squareness' of the current-voltage characteristics.

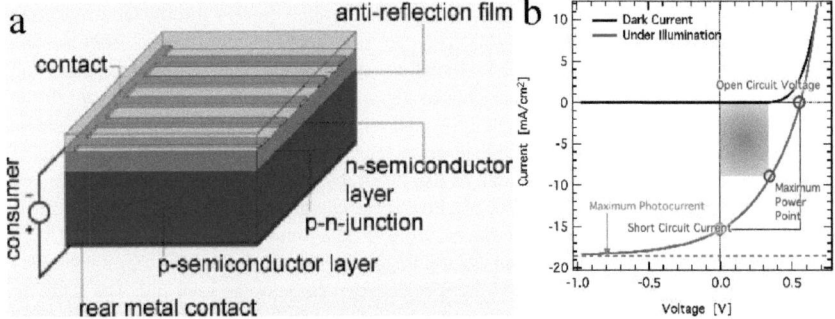

Fig. 2.1.: (a) Schematic of an inorganic p-n-junction solar cell and (b) current-voltage characteristics of organic solar cells, both modified from [4].

Organic solar cells are usually composed of four different materials. A transparent anode at the top to allow sunlight to reach the layers beneath. Indium tin oxide (ITO) is commonly used for this purpose. The cathode on the other side of the device is commonly made of aluminum. In between are a donor and an acceptor material. The donor is specifically designed to absorb sunlight and form tightly bound electron-hole pairs called excitons. The acceptor material is introduced to ease the separation of the neutral exciton into free charge carriers. The main difference between organic and inorganic solar cells is that in organic solar cells there is no space charge region and the separation of generated electron-hole pairs does not come freely. Instead a potential difference arising from different work functions of the anode and cathode materials is required for successful separation. Combining this with the fact that organic materials either do not form crystals at all or are less ordered than metals, since crystallinity is upheld by weak non-bonded interactions instead of strong covalent bonds, sheds a bad light on the efficiency of organic solar cells at first glance. However, in contrast to metals they may self-assemble and self-heal and their absorption length, i.e. the thickness of material required to absorb all incoming light, is three orders of magnitude lower ($\approx$ 100 nm) than that of silicon (300 μm). The steps involved in generating current from light in organic solar cells are as follows:

i. Light absorption and exciton generation
The incoming light passes through the transparent anode and is absorbed in the donor material, creating strongly bound electron-hole pairs, the excitons.

ii. Exciton diffusion to the acceptor interface
Due to the low dielectric constant and hence large screening length of organic materials, charges can see each other well and excitons are therefore strongly Coulomb bound. The electrically neutral exciton can only move by diffusion.

iii. Exciton dissociation and polaron pair generation
Excitons dissociate only at energetically favorable acceptor molecules, where the energy gain from the lowest unoccupied molecular orbital (LUMO) difference of donor and acceptor is larger than the exciton binding energy. If this is the case the electron transfers to the acceptor, leaving a hole in the donor. The resulting electron-hole pair is still Coulomb-bound and now called a geminate or polaron pair.

iv. Polaron pair dissociation leading to free electron-hole pairs
To separate the polaron pairs an electric field is required. It usually arises solely from the built-in voltage due to the work function difference (equalization of the Fermi levels) of anode and cathode material but may also be aided by an externally applied voltage.

v. Charge transport resulting in photo current
The now free charge carriers travel through the donor and acceptor material at a speed depending on their hole and electron mobilities toward the electrodes, where they are collected and finally yield the desired photocurrent.

Clearly, there remain many challenges in the design of efficient organic solar cells, some of which shall be highlighted below along with possible solutions. While the absorption coefficient is high, the absorption bands, i.e. the fraction of the spectrum of light that can be absorbed, tend to be narrow. Synthesis of new compounds and multijunction solar cells address this problem.

2. Organic Electronics

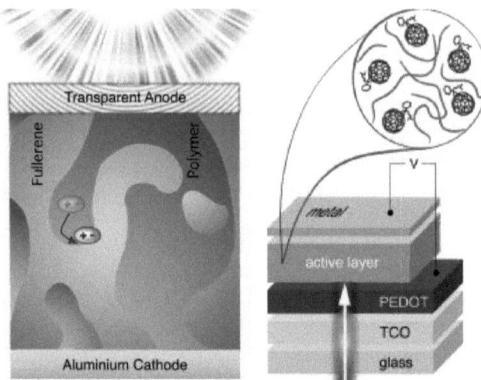

Fig. 2.2.: Schematic of exciton diffusion in a polymer-fullerene blend solar cell (left) and of the corresponding device structure (right). Taken from [4].

Also the exciton diffusion length prior to recombination is only a few nanometers. Bulk heterojunction solar cells, where donor and acceptor material are intermixed significantly reduce the distance an exciton has to travel to reach the donor-acceptor interface. They also introduce the problem of finding a percolation path toward the electrodes for the free charges, however. Preventing the recombination of polaron (geminate) pairs, one of the major loss mechanisms in solar cells, is still an unresolved issue.

2.1.2. Organic Light Emitting Diodes

Organic light-emitting diodes (OLEDs) do exactly the opposite of solar cells: they convert electrical current to light. The most simple setup for an OLED would be to place an organic light emitting material in between a high and a low work function metal. Electrons would enter the emission layer (EL) from cathode and holes from the anode. Upon recombination in the electroluminescent material they form a neutral exciton, which then decays to the ground state emitting a part of the liberated energy in the form of light. The color of the emitted light depends on the energy difference between the lowest unoccupied molecular orbital (LUMO) and the highest occupied molecular orbital (HOMO), since most decay processes will correspond to an electron from the LUMO filling a hole in the HOMO. In an oversimplified picture, the efficiency of the OLED should be good if the Fermi level of the anode were to match the HOMO and the Fermi level of the cathode the LUMO of the emitting layer. One of the electrodes must be transparent to allow for the transmission of light. The material of choice is usually indium tin oxide (ITO). With an ITO work function of 4.7 eV this limits the choices for the anode material to low work function metals such as Mg, Ca, Al or Li. Single layer devices are not exceptionally efficient, since they tend to inject more electrons than holes and also allow the charge carriers to cross from anode to cathode without forming excitons at all. To prevent this, multilayer devices are used in practice. There are different concepts containing all or some of the following additional layers:

- Hole injection layer (HIL)
  Hole injection layers may be used to smooth the surface contact with the ITO and increase performance and color stability of the OLED by providing a uniform conduction path for holes to cross the ITO/HIL interface [5].

- Hole transmitting (HTL) and electron transmitting layer (ETL)
  The transport layers provide intermediate energy states to allow holes (electrons) to cascade through smaller gaps in case there is a barrier between the metal work functions and the HOMO (LUMO) of the emission layer. They additionally pervent migration of contaminants from the electrodes to the emission layer (EL).

- Hole blocking (HBL) and electron blocking layer (EBL)
  These layers have very deep HOMO or high LUMO levels to prevent holes or electrons from passing through them and thereby preventing non-radiative decay pathways to the electrodes.

To illustrate the function especially of blocking and transporting layers a diagram displaying the Fermi levels of the electrodes along with the HOMO and LUMO levels of the organic materials in an OLED is shown in fig. 2.3 along with the complete device structure of a slightly different exemplary OLED. The main improvements that still have to be made to allow OLEDs to rival their inorganic counterparts are increasing the lifetime of the devices as well as the charge carrier mobility of the organic layers while a clear advantage is their flexibility.

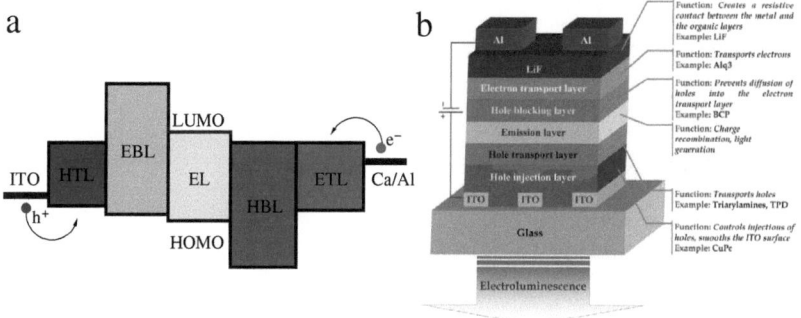

Fig. 2.3.: (a) Energy level diagram for a multilayer OLED illustrating the effect of blocking and transmitting layers. (b) Schematic of a complete OLED device, taken from [6].

2.1.3. Organic Field Effect Transistors

Organic field effect transistors (OFETs) are the basic building blocks for flexible integrated circuits and displays. The working principle of organic semi-conductors is identical to that of a metal-oxide-semiconductor field-effect transistor (MOS-FET), whose concept can be traced back to ideas of Julius Edgar Lilienfeld in 1930 [7]. An OFET is basicly a resistor that may be adjusted non-mechanically by application of an external voltage. It operates as a capacitor where one plate (the organic semiconductor) is the conducting channel between two ohmic contacts, termed source and drain. By applying a voltage to the second plate of the transistor, the gate electrode, which is separated from the organic semiconductor by an insulating layer (dielectric), the density of charge carriers in the channel and therefore the resistance between source and drain is modified. In the ideal case there is no conductance in the 'off' state. The 'on' state corresponds to sufficient voltage being applied between gate and source electrode to allow for charge transport in the channel. The current will initially increase linearly with the source-drain voltage until it finally saturates. Six components are required to realize such a device: a source, gate and drain electrode, an organic semiconductor, an isolator (dielectric) and a substrate [8]. The substrate is the basis upon which the rest of the OFET is built. It needs to meet the mechanical requirements of the desired application, such as flexibility, transparency and longevity, and provide good adhesion to the other components. Polyvinylchloride (PVC) or polyethyleneterephthalat (PET) foils are common examples of flexible substrates. If flexibility is not important, silicon wafers are also used. The most important property of the source, drain and gate electrodes is a high charge carrier mobility, so polymers such as polyethlenedioxythiophene (PETDOT) or polyalanine are possible candidates. In case of inorganic electrodes, gold (Ag) is commonly used, because its Fermi level closely matches the LUMO of many organic semiconductors. Another important aspect is to minimize the distance between source and drain, i.e. the length of the channel, since shorter channels correspond to higher efficency transistors. The channel itself forms at the dielectric-semiconductor interface. Hence the charge carrier mobility in the organic semiconductor is the key property of any OFET. It must be high enough to allow useful quantities of source-drain current to flow while being modulated by reasonable voltage and power. Also, the material must accept charges from the source electrode without a substantial barrier. Pentacene, tetracene, poly(3-hexylthiophene) (P3HT) and polyarylamine (PTAA) are possible materials. The dielectric layer is required to block off the semiconducting channel from the gate electrode. It must have an extremely high resistance and form close to defect-free layers. Examples are polymethylmethacrylate (PMMA) and polyvinylphenol (PVP).

The two performance parameters requiring optimization in an organic FET are the field effect mobility (μ_{FE}) and the on/off ratio [9, 10]. The mobility is related to the current in the 'on' state of the transistor. There are two different definitions for the current depending on whether the transistor is in the linear or in the saturated regime, since in the latter case it is independent of the source-drain voltage:

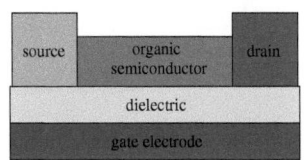

Fig. 2.4.: Schematic view of an organic field effect transistor.

$$I_{on} = [WC_i\mu_{FE}V_D(V_G - V_0)]/L$$
$$I_{on} = [WC_i\mu_{FE}(V_G - V_0)^2]/(2L) \qquad (2.1)$$

Here W and L are channel width and length, respectively, C_i is the capacitance per unit area of the gate dielectric, and V_D, V_G and V_0 are the drain-source, gate-source and threshold voltages, respectively. The on/off ratio is the quotient of this current with the current that flows in the absence of a gate field due to imperfections.

$$I_{on}/I_{off} = \mu_{FE} C_i V_G / (2\mu_r \rho h) \qquad (2.2)$$

where μ_r is the mobility and ρ the density of residual charge and h is the height of the semiconducting layer. To build well-functioning devices the on/off ratio must surpass 10^6. Both parameters must still be improved in OFETs, but the mobility is certainly the more challenging to lift OFETs out of the 'low cost - low performance' regime.

2.2. Measurements of Charge Mobility

Experimental measurements of the charge mobility are of great importance to gain an understanding of the influence of morphology and chemical structure on charge transport properties and to compare with computer simulation results. Below, we describe three different experimental methods to measure the mobility: time of flight, pulse radiolysis and field effect transistor (FET) measurements. The different techniques rarely yield the same order of magnitude for the mobility, since pulse radiolysis measures the local mobility and thus focuses on the neighboring pairs with the best coupling, while time of flight measures the bulk mobility heavily influenced by traps and defects and lastly FET measurements depend on the device structure and interfaces between different layers.

2.2.1. Time of Flight

Time of flight measurements monitor the time it takes for a packet of charge carries to travel through a μm thick sample. For this purpose the sample is placed between two non-injecting electrodes, which are used to apply electric fields to the sample. Both electrodes are semitransparent to allow photogeneration of charges by absorption of short pulses of laser light. Upon generation a packet of charge carriers travels through the medium from one electrode to the other. The monitored current reveals the transit time of the carriers allowing calculation of the mobility in the sample. Depending on the side chosen for photogeneration hole or electron mobilities may be measured.

For organic materials the chosen substrate is mostly indium tin oxide (ITO) with the counter electrode being made of aluminum. The current monitored shows three distinct features in the ideal, non-dispersive case (see Fig. 2.5): a strong initial peak, since charges are generated as electron-hole pairs and one carrier species immediately recombines at the respective electrode. The other travels at a constant velocity through the sample. This leads to a plateau region in the measured transient current. Once the packet of charge carriers has transversed the sample and reached the electrode on the other side, the current abruptly drops. The transit time t_T is given by the intersection of an extrapolation of the plateau region and the tangent to the tail of the transient. Once the transit time is known the mobility may be calculated from

$$\mu_{TOF} = \frac{d}{t_T E}, \qquad (2.3)$$

where d is the distance the charge carriers traveled, i.e. the thickness of the sample or the distance between electrodes, and E is the applied electric field.

2. Organic Electronics

Time of flight is one of the main techniques to measure mobilities in organic compounds [11, 12]. In case of organic or most real imperfect systems there are some difficulties involved. First, the spatial extension of the generated charge carrier packet must be significantly smaller than the sample size. This is of course governed by the penetration depth of the laser depending on the absorption of the sample. This difficulty has been significantly reduced by the charge-generation layer technique [13], where another thin, but highly absorbing layer is inserted between electrode and sample to guarantee a localized charge carrier generation. Nonetheless charge transport through organic materials is usually dispersive, i.e. charges will take different paths and hence different times to pass through the sample. This leads to a distribution of transit times and often makes it impossible to see a clear kink on linear scale. Therefore the transit time is determined on double logarithmic scale in case of dispersive transport, which significantly reduces the accuracy. Finally, the number of generated charges must be small enough not to distort the external electric field. This restricts the experiments to low excitation intensities and low electric fields on the order of $10^7 - 10^8$ V/m. Note that due to the size of the sample and the low charge carrier densities the time of flight technique measures the bulk mobility of the sample and is sensitive to disorder and defects within the sample [14, 15].

2.2.2. Pulse Radiolysis - Time Resolved Microwave Conductivity (PR-TRMC)

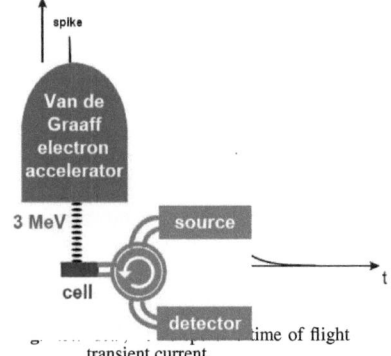

Fig. 2.6.: Experimental setup for a PR-TRMC measurement.

In PR-TRMC the mobility of a sample is calculated from the change in conductivity $\Delta\sigma$ in the sample measured by the corresponding change in microwave absorption. For this purpose the setup depicted in Fig. 2.6 is used. The sample is irradiated with single pulses of 3 MeV electrons from a Van de Graaff accelerator with a pulse duration between 0.2 and 250 ns. The penetration depth of the electron beam ($\approx$ 15 mm) must significantly exceed the sample size ($\leq$ 3 mm). The high energy of the incident electron beam is transferred to the medium by discrete excitation and ionization along its track. Microwaves in the GHz range are directed toward the sample by a waveguide. After propagating through the sample the waves are reflected at the end of the cell by a metal plate and directed toward the detector by a circulator. The decrease in power of the microwave allows calculation of the sum of electron and hole mobilities in the sample. The mathematical derivation is sketched based on [16, 17] below.

Consider a microwave traveling along the z-axis in a uniform waveguide containing a homogeneous medium. Its corresponding electric field is: $E(x, y, z) = E_0(x, y) \cdot e^{-\gamma z}$. The propagation constant γ is a complex number with the real part representing the change in amplitude due to attenuation and the imaginary part the change due to the wave nature of propagation. For a rectangular waveguide of width a, the propagation constant is given by $\gamma^2 = \left(\frac{\pi}{a}\right)^2 - \omega^2 \mu_0 \mu_r \epsilon_0 \epsilon_r$. Here ω is the radial frequency of the waves, μ_0 the permeability of the vacuum, μ_r the relative permeability of the medium, ϵ_0 the permittivity of vacuum and ϵ_r the relative dielectric constant of the medium $\epsilon_r = \epsilon' - i\epsilon''$, where ϵ'' takes into account all sources of loss in the medium. Taking the imaginary part of γ^2 to be a lot smaller than the real part we may write $\mathrm{Re}\{\gamma^2\} \approx \frac{\omega^2 \epsilon''}{2c^2\beta}$, where $c = (\mu_0\epsilon_0)^{-1/2}$ is the speed of light and μ_r is taken to be one. An increase in the conductivity of the medium $\Delta\sigma$ leads to a change in ϵ'' given by $\Delta\epsilon'' = \Delta\sigma/\omega\epsilon_0$ and therefore in the real part of the propagation constant. Thus one can measure a difference between incident power P_i and reflected power P_r depending on the conductivity of the medium:

$$\frac{\Delta P_r}{P_r} = \frac{(P_r/P_i)^{(\Delta\sigma)} - (P_r/P_i)^{(0)}}{(P_r/P_i)^{(0)}} \approx -A \cdot \Delta\sigma \tag{2.4}$$

The latter linear approximation holds for small absorptions $\Delta P_r/P_r$ and A is called the sensitivity factor. The change in conductivity of the sample is related to the mobility as follows:

$$\Delta\sigma = eN_p \sum \mu_{TRMC} \tag{2.5}$$

Here e is the elementary charge, N_p is the concentration of charge carriers, taken to be identical for negative and positive charges since the incoming electron beam generates electron-hole pairs, and $\sum \mu_{TRMC} = \mu_+ + \mu_-$ the mobility of the positive and the negative carriers. The concentration of the charge carriers may be calculated from the energy dose absorbed from the high-energy electron beam via $N_p = \frac{W_{eop}D}{eE_p}$. $W_{eop} < 1$ is the probability that the generated electron-hole pairs survive until the end of the pulse and hence the measurement, D is the energy dose deposited in the sample due to the electron pulse and E_p is the energy required to form an electron-hole pair. Taking these quantities to be known the mobility may be calculated as:

$$\sum \mu_{TRMC} = \frac{E_p}{W_{eop}D}\Delta\sigma \tag{2.6}$$

PR-TRMC has been applied to measure the mobility in columnar discotics [18, 19, 20, 21, 22] or (molecular) crystals [23]. In comparison to other measurement techniques, it measures the local mobility of small well-ordered domains, i.e. due to the ultrashort nanosecond timescale of the observations there is no time for charge carriers to diffuse to domain boundaries or impurity trapping sites. Defects and traps in the morphology are hence not seen by PR-TRMC and it is even attempted to measure single molecule mobilities with this technique. Therefore PR-TRMC measurements tend to yield higher mobilities than non-local methods.

2.2.3. Field effect transistor and diode measurements

Another approach to measure the mobility of an organic material is to include it as a functional layer in a device and extract the mobility from the device characteristics [24]. In a field effect transistor (FET) this is accomplished by measuring the current through the device in the linear or saturated regime and calculating the mobility from eqn. 2.1. Since the current flows through a very narrow channel at the dielectric to organic semiconductor interface, it is affected not only by defects within the organic layer, but also the surface topology and polarity of the dielectric, traps at the interface and the contact resistances at the source and drain metal/organic interfaces. The mobility is gate-voltage dependent since the first injected charges might have to fill defects. It also strongly dependens on charge carrier density, which in general is much higher in OFET measurements than in TOF or PR-TRMC. The last alternative method, which shall be mentioned here, is measurement of mobility in a diode configuration. Diodes are built simply by sandwiching an organic layer between two electrodes. The electrodes are chosen in such a way, that only electrons or holes are injected at low voltage. In the absence of traps and at low voltages U, the resulting current scales quadratically with applied bias voltage, since the current is space-charge limited. This means that the injected charges reach a maximum because their electrostatic potential prevents additional charges from being injected. The charge density is hence not uniform across the sample but largest near the injecting electrode. The current-voltage characteristics, from which the mobility μ may be extracted, may be expressed as:

$$I = \frac{9}{8}\epsilon_0\epsilon_r\mu\frac{U^2}{L^3} \qquad (2.7)$$

where ϵ_r denotes the dielectric constant of the medium and L is the device thickness. In the presence of traps and at high fields, additions must be made to the above formula.

3. Computer simulations

A many particle system, such as water, a membrane, an organic or liquid crystal, is completely described by the time-dependent many-body Schrödinger equation $i\hbar\frac{\partial}{\partial t}\Psi = \mathcal{H}\Psi$. Due to the incredibly large number of variables the solution thereof is impossible for any problem involving more than two simple molecules, even with sophisticated numerical approaches. Computer simulations in general and those of charge transport in particular thus face multiple time and length scale challenges. First, accurate knowledge of the molecular orbital structure of molecules is needed. Second, large systems must be treated to get accurate impressions of the morphologies and disorder therein. Third, a charge traveling over large distances by hopping between molecules, i.e. infrequent events, must be handled. All this can be done within certain approximations by a hierarchy of different methods: quantum chemical methods to obtain molecular orbitals, energy levels and transfer integrals, molecular dynamics to gain realistic morphologies and finally kinetic Monte Carlo to sample infrequent events as described in this chapter.

3.1. Ab initio quantum chemistry

Quantum Chemistry aims to solve the Schrödinger equation for the equilibrium configuration of a molecule to obtain its electronic structure in the ground and excited states. Therefore the time-independent Schrödinger equation is solved by numerical methods focusing on the electronic wave functions [25].

Ab initio (lat.: from the beginning) methods are quantum chemical methods solving the Schrödinger equation using solely physical constants and no experimental knowledge. The term was coined to distinguish them from semi-empirical methods, requiring constants obtained by fitting to experimental data. In the late 1920s D.R. Hartree developed a self-consistent field (SCF) procedure to calculate approximate wavefunctions and energies for atoms and ions to become known as the Hartree method. In 1928 J.C. Slater and J.A. Gaunt showed that the method could be based on solid mathematical theory applying the variational principle to a trial wavefunction written as a product of single-particle functions. In 1930 V.A. Fock and J.C. Slater corrected the Hartree method to respect the antisymmetry of the wave function and the Pauli exclusion principle giving rise to the Hartree-Fock (HF) method. Due to the computational demands of the HF method it wasn't until the 1950s and the invention of electronic computers that it became widely used. In the following years, methods reducing the HF approximations, especially no longer neglecting the electron correlation, were introduced. They are known as Post-Hartree-Fock methods and include methods such as perturbation theory of nth order, configuration interaction (CI) and coupled cluster (CC). The main proofs underlying ab initio theory as well as technical details such as basis sets for the representation of orbitals and methods to numerically solve the corresponding matrix equations are outlined in appendix A.1.

3.2. Density Functional Theory (DFT)

Density functional theory (DFT) is an alternative to ab initio methods for solving the non-relativistic, time-independent Schrödinger equation $\mathcal{H}|\Phi\rangle = E|\Phi\rangle$ [26]. Since the ab initio approach splits the wave function of the system into single electron wave functions, for a system with N electrons this means there are $3N$ continuous variables to be fitted. Assuming that $3 \leq p \leq 10$ parameters per variable are required to yield a fit of reasonable accuracy, this means that p^{3N} parameters must be optimized for an N electron system. As was shown in the ab initio section, it is not quite that bad, but nonetheless the exponential wall exists and limits traditional wave function methods to molecules with a small number of chemically active electrons: $N \approx O(10)$ [27]. This problem is overcome by DFT, which is expressed in terms of the density $n(\mathbf{r})$ in the Hohenberg-Kohn formulation and in terms of $n(\mathbf{r})$ and the single-particle wave functions $\psi_j(\mathbf{r})$ in the Kohn-Sham formulation. The proof of the Hohenberg Kohn theorem, details of the Kohn-Sham formulation and different DFT functionals are presented in appendix A.2.

3.3. Molecular dynamics simulations

Molecular dynamics (MD) simulations focus on the properties of systems containing a large number of molecules. Therefore it is assumed that the electrons follow the motion of the nuclei instantaneously, since they are much faster and lighter. As a result, it suffices to describe only the relatively slow motion of the heavy nuclei by means of classical mechanics, i.e. Newton's equation of motion $\mathbf{F} = m\mathbf{a}$. Quantum mechanical effects are included in the force field, which describes how atoms interact with each other and defines the potential governing the system.

3.3.1. Force fields

The term force field is slightly misleading, since it refers to the parameters of the potential used to calculate the required forces (by taking the gradient) in MD simulations. The underlying idea is to create a certain number of atom types based upon which any bonds, angles, dihedrals and long-range interactions may be described. More atom types than elements are necessary, since the chemical surroundings greatly influence the parameters. For example the angle between carbons in an alkyl side chain clearly differs from that of carbons in a phenyl ring, so two different atom types must be used to describe them.

The force field we decided to use throughout our work is OPLS (Optimized Potential for Liquid Simulations), developed by William L. Jorgensen at Purdue University and later at Yale University [28, 29, 30, 31, 32, 33]. The functional form of OPLS is based upon AMBER (Assisted Model Building and Energy Refinement), originally developed by Peter Kollman's group at the University of California San Francisco [34, 35, 36, 37], and looks as follows:

$$V(\{\mathbf{r}_N\}) = \sum_{bonds} \frac{1}{2} k_b (r - r_0)^2 + \sum_{angles} k_\theta (\theta - \theta_0)^2$$
$$+ \sum_{torsions} \left\{ \frac{V_1}{2}[1 + \cos(\phi)] + \frac{V_2}{2}[1 - \cos(2\phi)] + \frac{V_3}{2}[1 + \cos(3\phi)] + \frac{V_4}{2}[1 - \cos(4\phi)] \right\}$$
$$+ \sum_{i=1}^{N-1} \sum_{j=i+1}^{N} \left\{ 4\epsilon_{ij} \left[\left(\frac{\sigma_{ij}}{r_{ij}}\right)^{12} - \left(\frac{\sigma_{ij}}{r_{ij}}\right)^6 \right] + \frac{q_i q_j}{r_{ij}} \right\} f_{ij}$$

Bonds and angles are described by harmonic potentials, i.e. springs, since they are very strong and show only zero point fluctuations at room temperature. The quantum chemical vibrational spectra of the molecules are lost in such a description, which makes it useful only for large molecules, where only the conformations of the entire molecule are of interest. The dihedral potential is described by a cosine expansion and may take any value within 360° depending on the height of the barrier between the low energy conformations. This makes the precision of the dihedral potential barrier crucial for many polymer properties. Dihedral potentials always possess a symmetry around 180°. The long range interactions are only counted for atoms three or more bonds apart. They consist of Coulomb and Lennard Jones two-body interaction terms. The Lennard Jones potential is a combination of attractive van der Waals forces due to induced dipole-dipole interactions, interactions between permanent dipoles and empirical repulsive forces due to Pauli repulsion. The scaling factor f_{ij} equals 0.5 for 1-4 interactions, i.e. interactions between a given atom (1) and one connected by three consecutive bonds (4) which are additionally described by a dihedral potential, and one otherwise. It is important to note that since the bonds never break due to their functional form, MD simulations based on the OPLS force field are not suitable to describe any kind of chemical reaction. Also, for OPLS correlations between bonds and angles are neglected. Force fields such as MM3 [38] and MM4 [39] take such correlations into account.

3.3.2. Calculation of force field parameters

Force field parameters are the constants describing the strength of the harmonic potentials of bonds, the barrier heights defining dihedral rotation angles and the strength of long range interactions. Since they should not be taken for granted and may not exist in established force fields for certain compounds, the following outlines how they are derived [36].
In general any force field derivation must start with very small molecules, such as CH_3, CH_4, or maybe benzene, to limit the number of variables. Larger molecules can only be considered once the force field parameters for most constituents are already known. This is the case for all molecules of interest to us and so only a few parameters had to be calculated while most were chosen in analogy to already parameterized compounds.

Bonds and angles

The equilibrium values for bonds as well as angles are taken from x-ray data. The values for the force constants are derived by fitting to experimental vibrational frequency data. While one should actually go through a more complex procedure to ensure that the geometries of simple molecules match experimental data as good as possible after energy minimization, it is believed that in most cases the difference is negligible.

Dihedrals and impropers

There are two different approaches as to how to calculate dihedral potentials. One is to optimize the dihedral potential for the simplest possible molecule and then apply it to larger ones containing the same dihedral. The other is to optimize the dihedral parameters to best describe a large number of different molecules. Note that, while the latter method might sound more accurate at first, it also leads to a dependence on the set of chosen molecules. In either case, the dihedral parameters are computed based on ab initio or DFT methods as follows [40]. The described methodology only works in case of one unknown parameter, otherwise multidimensional fitting becomes necessary.

  i. **Ab initio calculations**
     - Scan dihedral (or improper) of interest
     - Optimize geometry at each step
     - Calculate change in potential energy

For this purpose, either perturbation (MP2), restricted Hartree-Fock (RHF) or hybrid methods between Hartree-Fock and density functional theory (B3LYP) are used. The basis set chosen for the geometry optimization is 6-311G(d,p) unless specified otherwise.

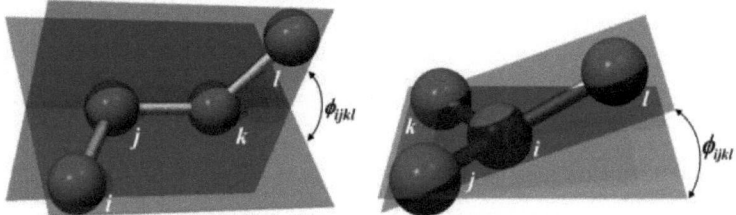

Fig. 3.1.: Dihedral and improper angle between atoms i,j,k,l. Taken from [41].

 ii. **Potential energy according to MD**
     - Set parameters of dihedral in question to zero
     - Minimize energy of configurations
     - Compute potential energy of each configuration

This requires prior knowledge of all other force field parameters for the molecule. The energy of the configurations is minimized using the steepest descent algorithm, which follows the potential energy surface defined by the MD force field along its steepest gradient and thereby moves the atoms into the closest local minima.

iii. **Fitting of the parameters**

- Subtract MD from ab initio results
- Obtain parameters from fitting a dihedral function to the resulting curve

The difference between ab initio and MD energies corresponds exactly to the influence of the dihedral, since the dihedral parameters were set to zero in the MD calculation and all other interactions are already included in the force field. The mathematical representations used for dihedrals in this work are:

- Proper dihedrals

$$V(\{r\}) = C_n \cdot (1 + \cos(n\phi))$$

- Ryckaert-Belleman dihedrals

$$V(\{r\}) = \sum_{n=0}^{5} C_n \cos^n(\psi)$$

where $\phi = \psi - 180°$.

iv. **Checking the results**
- Rerun MD simulations using new parameters
- Compare MD and ab initio energies

Lennard-Jones parameters

Lennard-Jones parameters are the most difficult to derive, since they have to be adjusted in order to match experimentally measured bulk properties. This is also where the advantage of OPLS over other force fields comes into play when dealing with organic molecules. While most bond, angle and dihedral parameters were taken from the force field developed by Weiner et al. [34] the Lennard Jones parameters were calculated by taking the solvent molecules into account explicitly when necessary. This is achieved by carrying out MD or Monte Carlo (MC) simulations of organic liquids, e.g. CH_4 or C_2H_6, and then empirically adjusting the Lennard Jones (σ and ϵ) parameters to match the experimental densities and enthalpies of vaporization. A difficult issue is the factor to scale down the Lennard Jones 1-4 interactions (interactions between molecules only three bonds apart). It has to be done, since otherwise the r^{-12} term would lead to unphysically high repulsions. The value chosen is somewhat arbitrary and force field dependent, however. The same holds true for the combination rules of Lennard Jones parameters, i.e. the choice of LJ parameters describing the interaction between two different atom types. There are two main competing methods:

- Lorentz-Berthelot

$$\sigma_{ij} = \frac{1}{2}(\sigma_{ii} + \sigma_{jj})$$
$$\epsilon_{ij} = (\epsilon_{ii}\epsilon_{jj})^{1/2}$$

- Geometrical average

$$\sigma_{ij} = (\sigma_{ii}\sigma_{jj})^{1/2}$$
$$\epsilon_{ij} = (\epsilon_{ii}\epsilon_{jj})^{1/2}$$

Each force field requires a certain combination rule to be used. In case of OPLS this is the geometrical average.

Partial Charges

Each atom in a molecule may carry partial charge because of the rearrangement of the outer-shell electrons due to its surroundings. This quantity is also fixed during simulation time in standard, non-polarizable force fields. The charge calculations are therefore based on the optimized geometry of a single molecule in vacuum, which we obtained via B3LYP with 6-311G(d,p) basis set. The charges may then be obtained by fitting to the electrostatic potential, i.e. adjusting the partial charges at the centers of the nuclei in such a fashion that the electrostatic potential given by the wave functions is best reproduced. An example of such a method is CHELPG (CHarges from ELectrostatic Potentials using a Grid based method) [42]. Charge calculations are done using higher-level methods and basis sets than the geometry optimization, e.g. B3LYP/cc-pVTZ. There are two known problems with this straight-forward approach however: Firstly, oftentimes considerable variation is seen when charges are computed for different conformations of a molecule, which is especially problematic for molecules with multiple low energy conformations such as propylamine [35]. Secondly, standard electrostatic potential (ESP) charges tend to miscalculate charges of "buried" atoms, since they are in general far away from the surface points at which the ESP is evaluated and close to differently charged atoms. This leaves them statistically underdetermined and they may vary greatly when attempting to improve the quality of the least-square ESP fit. For these reasons, the restricted ESP charge model was developed [43]. It consist of a least-squares fit of the charges to the electrostatic potential (as before), but with hyperbolic restraints on heavy atom charges. This is followed by a second fitting stage, needed to fit methyl groups which require equivalent charges on hydrogen atoms which are not equivalent by molecular symmetry. Nonetheless, we decided to use CHELPG for our partial charge calculations, confirming the results by checking convergence of partial charge with increasing size of the basis set. For this purpose we start with charge calculations using the 3-21G(d) basis set and increase the basis set up to 6-311++G(3df,3pd) with five or six intermediate steps (see fig. 3.2).

Force field validation

There are multiple ways to test the validity of a force field, depending also upon what type of system is simulated. For proteins and other complex molecules, it is important that the Ramachandran plots, i.e. statistical distributions of consecutive dihedral angles along the backbone, correspond to those obtained experimentally to ensure sampling of the correct secondary structures in simulations. For liquids, the radial distribution function and higher correlations as well as the density should be correctly reproduced. The radius of gyration and the persistence length are checked for polymer chains. Crystalline compounds must be stable when simulated

3.3 Molecular dynamics simulations

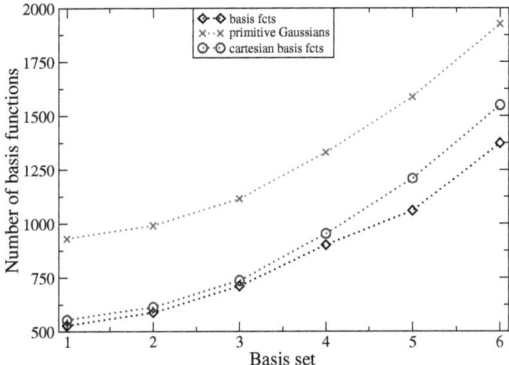

Fig. 3.2.: Comparison of basis set sizes for some of the Pople basis sets used, i.e. numbers one through six refer to 6-311G(d), 6-311G(d,p), 6-311++G(d,p), 6-311++G(2d,2p), 6-311G(2df,2pd) and 6-311++G(3df,3pd)).

starting from an x-ray crystallography structure and reproduce a similar melting point as experiments.

3.3.3. Integration schemes

MD simulations are based on classical mechanics and therefore aim at solving Hamilton's equations of motion.

$$\dot{q}_i = \frac{\partial \mathcal{H}}{\partial p_i} \quad (3.1)$$

$$\dot{p}_i = -\frac{\partial \mathcal{H}}{\partial q_i} \quad (3.2)$$

Here q_i and p_i are the generalized coordinates and momenta, respectively, and $\mathcal{H}$ is the classical Hamiltonian corresponding to the system. The index i labels the different particles (atoms, molecules,$\cdots$) in the system. Due to the Liouville theorem, which states that the phase space volume of a set of systems is invariant under time evolution, the following must hold for any distribution function $f(q_i, p_i, t)$ in phase space

$$\frac{d}{dt}f(\{q\},\{p\},t) = 0 = \frac{\partial f}{\partial t} + \sum_i \frac{\partial f}{\partial q_i}\dot{q}_i + \sum_i \frac{\partial f}{\partial p_i}\dot{p}_i := \frac{\partial f}{\partial t} + \{f, \mathcal{H}\}$$

where $\{f, \mathcal{H}\}$ is the Poisson bracket. One can now continue to define the Liouville operator $\mathcal{L}$ via $i\mathcal{L}f := \{f, \mathcal{H}\}$, i being the imaginary unit. Then the time evolution of any distribution in phase space, which is not explicitly time-dependent (as is the case for distributions of atoms) is described by the equation $\frac{d}{dt}f = i\mathcal{L}f$, which given an initial state at $t = 0$ has the formal solution

$$f(\{q(t)\}, \{p(t)\}) = e^{i\mathcal{L}t} f(\{q(0)\}, \{p(0)\}) \quad (3.3)$$

To gain a better understanding of how the Liouville operator acts, it is helpful to define $i\mathcal{L}_q := \sum_i \dot{q}_i \frac{\partial}{\partial q_i}$ and $i\mathcal{L}_p := \sum_i \dot{p}_i \frac{\partial}{\partial p_i}$. This allows us to calculate:

$$\begin{aligned} f_q(t) &= e^{i\mathcal{L}_q t} f(0) = e^{\sum_i \dot{q}_i t \frac{\partial}{\partial q_i}} f(0) \\ &= \prod_{i=1}^{N} e^{\dot{q}_i t \frac{\partial}{\partial q_i}} f(0) = \prod_{i=1}^{N} \left[\sum_{n=0}^{\infty} \frac{1}{n!} (\dot{q}_i t)^n \left(\frac{\partial}{\partial q_i} \right)^n \right] f(0) \\ &= f(q_1(0) + \dot{q}_1 t, q_2(0) + \dot{q}_2 t, \cdots, q_N(0) + \dot{q}_N t; p_1(0), \cdots, p_N(0)) \end{aligned} \quad (3.4)$$

So $\mathcal{L}_q$ is the generator of translations in the coordinates and $\mathcal{L}_p$ the generator of translations in the momenta. Since the two generators do not commute one has to apply the Trotter formula to express the full Liouville operator with them, i.e.

$$e^{A+B} = \lim_{n \to \infty} \left(e^{\frac{A}{2n}} e^{\frac{B}{n}} e^{\frac{A}{2n}} \right)^n \quad (3.5)$$

$$\Rightarrow e^{i\mathcal{L}t} \approx \left(e^{i\frac{\Delta t}{2} \mathcal{L}_p} e^{i\Delta t \mathcal{L}_q} e^{i\frac{\Delta t}{2} \mathcal{L}_p} \right)^n e^{O(\frac{\Delta t}{t})^2} \quad (3.6)$$

So after n translations by Δt the error is of the order of $(\Delta t)^2$. The coordinates and momenta of a system after a time step Δt may be computed using the Verlet integration scheme [44]:

$$q_i(t + \Delta t) = q_i(t) + \frac{p_i(t)}{m} \Delta t + \frac{1}{2m} F(\{q(t)\})(\Delta t)^2 \quad (3.7)$$

$$p_i(t + \Delta t) = p_i(t) + \frac{1}{2} F(\{q(t)\}) \Delta t + \frac{1}{2} F(\{q(t + \Delta t)\}) \Delta t \quad (3.8)$$

These equations approximate the originally continuous dynamics given by eqn. 3.3 within a truncation error of $(\Delta t)^4$. For application in MD simulations the algorithm was rewritten in terms of velocity $v := p/m$. An algorithm that updates the positions and velocities in the system using forces determined only by the positions at time t (not requiring calculation of forces at $t+\Delta t$) is the leap-frog algorithm [45], which is the algorithm of choice in the simulation package Gromacs [46] we use for MD simulations, and consists of the following two equations:

$$v(t + \frac{\Delta t}{2}) = v(t - \frac{\Delta t}{2}) + \frac{F(t)}{m} \Delta t \quad (3.9)$$

$$r(t + \Delta t) = r(t) + v(t + \frac{\Delta t}{2}) \Delta t \quad (3.10)$$

3.3.4. Thermostats

Molecular dynamics performed using solely the numerical integration schemes explained above corresponds to running in an NVE or microcanonical ensemble, i.e. keeping the number of particles N, the volume of the system V and the total energy E constant. There are two main reasons why it is often not desirable to use this ensemble. First, experiments cannot be carried out under such conditions, since it is impossible to control the total energy of a system. Second, numerical integration errors may lead to steady drifts in energy or the system may heat up due to external or frictional forces. To avoid these problems, simulations may be run using thermostats, which ensure constant temperature within the system [47]. The temperature in MD is given by the particles velocities

$$\frac{k_B T}{2}(3N - N_c) = \sum_{i=1}^{N} \frac{1}{2} m_i |\mathbf{v}_i|^2 \qquad (3.11)$$

where k_B is the Boltzmann factor, T the temperature, N the number of particles, m_i the mass of a single particle and $\mathbf{v}_i$ the corresponding velocity. The term $N_{df} = (3N - N_c)$ represents the degrees of freedom. Each particle may move in the x-, y- and z-direction, which leads to $3N$ degrees of freedom minus the number of conserved quantities N_c. In general, momentum $\mathbf{p}$ and angular momentum $\mathbf{L}$ are conserved. However, for periodic boundary conditions conservation of angular momentum no longer holds. Since the velocities of each particle are known at every time step in MD simulations, the obvious way to control the temperature (eqn. 3.11) is by scaling the velocity of each particle by a factor of λ to give the desired temperature. Let the temperature at time t be $T(t)$, then the change in temperature due to a multiplication of all velocities by λ is given by:

$$\Delta T = \sum_{i=1}^{} \frac{m_i (\lambda v_i)^2}{N_{df} k_B} - \sum_{i=1}^{} \frac{m_i v_i^2}{N_{df} k_B} = (\lambda^2 - 1) T(t) \qquad (3.12)$$

Thus a straightforward way to control the temperature is to multiply all velocities by a factor of $\lambda = \sqrt{T_0/T(t)}$ at every step. However, this does not allow for temperature fluctuations and hence does not sample a thermodynamic ensemble. Also since the algorithm acts solely on the kinetic energy, drifts in the potential energy are still possible and may lead to large discrepancies between potential and kinetic energy after a while.

Berendsen temperature coupling

A slightly weaker formulation of velocity scaling leads to the Berendsen thermostat [48], where the systems kinetics are coupled to an external heat bath with a given temperature T_0. Thereby the algorithm slowly corrects deviations of the system temperature $T(t)$ from the desired temperature T_0 by

$$\frac{dT}{dt} = \frac{1}{\tau}(T_0 - T(t)) \Rightarrow \Delta T = \frac{\delta t}{\tau}(T_0 - T(t)) \qquad (3.13)$$

i.e. the temperature deviation decays exponentially with a time constant τ. The factor to rescale the velocities is time-dependent and given by:

$$\lambda^2 = 1 + \frac{\delta t}{\tau_T}\left[\frac{T_0}{T(t-\delta t/2)} - 1\right] \qquad (3.14)$$

where the $\delta t/2$ is due to the leap-frog algorithm (eqn. 3.9). The factor τ_T is not identical with the time constant τ of the temperature coupling from eqn. 3.13, because the change in energy due to rescaling the velocities is partly redistributed between kinetic and potential energy, so that the change in temperature is less than the scaling energy. The precise relation between τ and τ_T is

$$\tau = 2C_V \tau_T / N_{df} k_B \qquad (3.15)$$

where C_V is the total heat capacity of the system. The ratio τ/τ_T ranges from 1 (gas) to 2 (harmonic solid) to 3 (water). In MD simulations the 'temperature coupling time constant' always refers to τ_T. In practice the scaling factor may only be chosen in the range $0.8 < \lambda < 1.2$ to avoid scaling by very large numbers, which may crash the simulations. Since the Berendsen thermostat still suppresses kinetic energy fluctuation, it does not generate a true canonical ensemble. Therefore, it is favored for equilibration runs due to the exponentially fast coupling, but not for production runs.

Nosé-Hoover temperature coupling

The Nosé-Hoover coupling was specifically developed to sample the correct canonical ensemble [49, 50]. For this purpose it treats the heat bath as an integral part of the system by introducing a thermal reservoir and a friction coefficient in the equations of motion.

$$\frac{d^2\mathbf{r}_i}{dt^2} = \frac{\mathbf{F}_i}{m_i} - \xi\frac{d\mathbf{r}_i}{dt} \qquad (3.16)$$

The equation of motion for the heat bath parameter ξ is

$$\frac{d\xi}{dt} = \frac{1}{Q}(T(t) - T_0) \qquad (3.17)$$

As before T_0 is the desired temperature and $T(t)$ the current temperature. The reference temperature and the constant Q determine the strength of the coupling. The latter is also called the 'mass parameter' of the reservoir. This approach does not give an exponential but an oscillatory relaxation. The period τ_T of the oscillations is given by

$$Q = \frac{\tau_T^2 T_0}{4\pi^2} \qquad (3.18)$$

The actual time the system takes to relax is several times larger than the chosen period of oscillation. Note that in the analysis of a trajectory with very frequent print-outs, as is required for example to compute the autocorrelation of transfer integrals (20 fs between print-outs), the fluctuations are clearly visible in the energy and lead to unwanted correlations. This lead us to choose either the velocity rescaling thermostat or use the NVE ensemble in such cases.

Velocity rescaling thermostat

To combine the advantages of the Berendsen, i.e. first order decay of temperature and no oscillations, and the Nosé-Hoover thermostat, i.e. sampling of a correct canonical ensemble, the velocity rescaling thermostat was developed [51]. It is identical to the Berendsen thermostat with an additional stochastic term allowing fluctuations of the kinetic energy K:

$$dE_{kin} = (K_0 - K)\frac{dt}{\tau_T} + 2\sqrt{\frac{KK_0}{N_{df}}}\frac{dW}{\sqrt{\tau_T}} \qquad (3.19)$$

Here N_{df} is the number of degrees of freedom and dW is a Wiener process. Other than a random seed for the Wiener process, there are no additional parameters. After its implementation in *Gromacs* 4.0 it became our choice for production runs.

Stochastic dynamics (SD)

The stochastic dynamics thermostat [52] takes a different approach from all thermostats described above by locally coupling every single particle to a heat bath such that

$$\frac{d\mathbf{r}_i}{dt} = \frac{\mathbf{p}_i}{m_i} \qquad (3.20)$$

$$\frac{d\mathbf{p}_i}{dt} = \mathbf{F}_i - \zeta\frac{\mathbf{p}_i}{m_i} + \mathbf{f}_i, \qquad (3.21)$$

where ζ is a constant friction parameter and the stochastic forces $\mathbf{f}_i$ have a mean of zero. The local scale thermostatting prohibits propagation of numerical instabilities arising from inaccurate calculations of collision-like processes, which allows use of a larger integration time step. However, the damping term violates Galilei invariance and thus violates momentum conservation visible in a damping of hydrodynamic correlations.

Dissipative particle dynamics (DPD)

The DPD thermostat combines the positive aspects of SD, such as numerical stabilization, with the advantages of Nosé-Hoover, like strict Galilean invariance and correct hydrodynamics [53]. It achieves this by dampening the velocity differences of nearby particles instead of the absolute particle velocities as is the case in SD. The stochastic forces also act on pairs of nearby particles, thus strictly fulfilling Newton's third law. The disadvantage is instantaneous heat transfer due to the locality of the thermostat, which may be seen as limit of infinite thermal conductivity. In the formulation of Español and Warren [54] the DPD equations read:

$$\frac{d\mathbf{p}_i}{dt} = \mathbf{F}_i - \zeta\omega^D(r_{ij})(\hat{r}_{ij} \cdot \mathbf{v}_{ij})\hat{r}_{ij} + \sigma\omega^R(r_{ij})\theta_{ij}\hat{r}_{ij} \qquad (3.22)$$

Here $\mathbf{v}_{ij} = \mathbf{v}_i - \mathbf{v}_j$ is the relative velocity between particle i and j, $\mathbf{r}_{ij}$ is the connecting vector and $\hat{r}_{ij}$ the unit vector in the same direction. ζ is the friction constant and σ the noise strength. ω^D and ω^R are r-dependent weight functions vanishing beyond a given cut-off. θ_{ij} is a Gaussian white noise variable with a first moment of zero.

3.3.5. Barostats

To equilibrate artificial starting configurations, it is important to run in the *NPT* ensemble and allow fluctuations in box size to keep the pressure and reach the desired density. The pressure tensor may be computed from quantities easily accessible in MD simulations by application of the Clausius virial theorem

$$\mathbf{P} = \frac{2}{V}(E_{kin} - \Xi) \tag{3.23}$$

where V is the volume of the simulation box, E_{kin} is the kinetic energy and Ξ is the inner virial tensor given by

$$\Xi = -\frac{1}{2}\sum_{i<j} \mathbf{r}_{ij}\mathbf{F}_{ij} \tag{3.24}$$

where r_{ij} is the vector connecting the particles i and j and F_{ij} is the force acting between them. All pressure correction algorithms work by changing the inner virial Ξ by scaling the inter particle distances. In case of an isotropic system the isotropic pressure may be calculated from the trace of the pressure tensor by

$$P = \frac{1}{3}\text{Tr}\{\mathbf{P}\} \tag{3.25}$$

Berendsen pressure coupling

In the Berendsen pressure coupling algorithm [48], similarly to the temperature coupling algorithm (sec. 3.3.4), the coordinates as well as the box vectors are rescaled at every step with a matrix μ. This leads to a first-order kinetic relaxation of the pressure towards the reference pressure $\mathbf{P}_0$

$$\frac{d\mathbf{P}}{dt} = \frac{\mathbf{P}_0 - \mathbf{P}}{\tau_p} \tag{3.26}$$

The scaling matrix μ is given by

$$\mu_{ij} = \delta_{ij} - \frac{\Delta t}{3\tau_p}\beta_{ij}\left[P_{0ij} - P_{ij}(t)\right] \tag{3.27}$$

Here β is the isothermal compressibility of the system, usually given by a diagonal matrix with identical elements on the diagonal, whose value is generally not known. This is not a severe problem, since it only influences the time-constant of the relaxation and not the average pressure itself. However, it only allows for changes in the length not the orientation of the box vectors. Same as for the thermostat, the weak-coupling scheme of Berendsen does not correctly describe a thermodynamical ensemble, but is ideal to equilibrate a system. Since all our production runs use NVT or NVE ensemble and we have used solely the Berendsen barostat for equilibration, we refrain from giving details on other methods. We would like to mention, however, the Parinello-Rahman barostat [55], which explicitely solves a matrix equation of motion for the box vectors to allow changes in the box shape. This is required for equilibration of crystalline systems in case the unit cell vectors are unknown. It also simulates an exact ensemble as does the Nosé-Hoover thermostat for temperature coupling.

3.4. Kinetic Monte Carlo (KMC)

For the analysis and simulation of atomic scale systems molecular dynamics is extremely popular, since it gives an accurate representation of the real physical system unless quantum dynamical effects are important or the Born-Oppenheimer approximation does not hold. The enormous drawback of MD methods is, however, that an accurate integration of the system requires timesteps capable of capturing even hydrogen vibrations, i.e. on the order of 10^{-15} seconds. While united atom and coarse-graining approaches somewhat remedy this issue, the maximum simulation time is still limited to less than a microsecond for most systems. This is called the time-scale problem. For systems were only rare events are of interest and transition rates between these events are known, there is an alternative simulation method which elegantly circumvents this problem: kinetic Monte Carlo (KMC) [56]. Monte Carlo methods have earned their name since they rely on the generation of random numbers as does the cashflow in the casinos of Monte Carlo, Monaco. Monte Carlo methods emerged in the late 1940's as electronic computers came to be used and were applied to solving integrals numerically by Monte Carlo integration [57] or sampling different geometries in physical ensembles by the Metropolis algorithm [58]. In the 1960's such methods were also developed for evolving systems dynamically from state to state. One of the first applications to atomistic systems was Beeler's simulation of radiation damage in 1966 [59]. In the 1990's the term kinetic Monte Carlo was coined for such methods. Today they are widely used in numerous different fields such as radiation damage, surface adsorption, diffusion and growth, statistical physics and last but not least of course charge transport.

3.4.1. Infrequent-event systems

The underlying premise is that the dynamics of the system of interest are characterized by occasional transitions from one state to another with long periods of inactivity inbetween. The key property of such systems is that, because it stays in one state for such a long time, it eventually forgets how it got there. Thus the rate constants k_{ij}, which characterize the time it takes for the system to escape from state i and move to state j, are independent of what state proceded i. This independence of transition probabilities from the history of previous transitions is the defining property of a Markov chain. State-to-state dynamics in such a system are called a Markov walk. Because the transition rate out of state i is given by the sum of rates to all surrounding states k_{ij}, a simple stochastic procedure may be designed to propagate the system correctly from state to state. If we possess exact knowledge of all rate constants governing the system, the state-to-state trajectory of such a simulation will be indistinguishable from that obtained by an MD simulation, i.e. the probability to observe a certain sequence of transitions will be identical in both cases, but the KMC result comes at far lower cost.

Rate constants

Because the system has no memory of its past at any given time it has the same probability of escaping from its current state to another throughout. It is thus governed by a first-order process with exponential decay statistics, i.e. the probability that the system has not yet left its current state i is given by

$$p_{survival} = \exp\{-k_{tot}t\} \tag{3.28}$$

where k_{tot} is the total escape rate for escape from that state. More interesting to us, however, is the probability $p(t)$ for the time of first escape from the state. Clearly, the integral of $p(t)$ at a given time t' is equal to $1 - p_{survival}(t')$. Thus, the negative time derivative of $p_{survival}$ gives the probability distribution function for the time of first escape

$$p(t) = k_{tot}\exp\{-k_{tot}t\} \tag{3.29}$$

The average time of escape τ is therefore the first moment of the above distribution, $\tau = \int_0^\infty tp(t)dt = \frac{1}{k_{tot}}$. Since escape may occur along any number of different routes and the system has a fixed probability per time of finding each route, the total escape rate must be the sum of the individual rates

$$k_{tot} = \sum_j k_{ij} \tag{3.30}$$

In other words for each escape way there is again an exponential first-escape time distribution.

$$p_{ij}(t) = k_{ij}\exp\{-k_{ij}t\} \tag{3.31}$$

Obviously, only one of them may be the first to occur.

Exponentially distributed random numbers

To deal with the above process in a Monte Carlo fashion, it is important to generate a random number t_{draw} from an exponential distribution $p(t) = k\exp\{-kt\}$. To this end, a random number r is drawn in the (0,1) interval. Multiple random number generators exist to achieve this. Given r we obtain t_{draw} by:

$$t_{draw} = -(1/k)\ln\{r\} \tag{3.32}$$

An important thing to note is that it does not matter whether the random number is drawn from the $0 < r < 1$ or the $0 < r \leq 1$ interval, but in case zero is part of the range this might cause an ill-defined $\ln\{0\}$ operation, which must be avoided.

3.4.2. The KMC procedure

Let us assume that a system begins in state i and the set of rate constants k_{ij} for all pathways are known. A random number t_j is then drawn from the exponential distribution of first-escape times for each pathway leading away from i to j. Since the system must decide for one escape route, the pathway j_{min} with the lowest value of t_j, i.e. t_{min}, is chosen. The system is then moved to the state j_{min}, the other drawn times are discarded and the time of the overall system is advanced by t_{min}. The same procedure then repeats for the new state. This algorithm is not efficient, because a random number is drawn for each of the j escape paths, although all except the one chosen are discarded. A far more efficient way to implement KMC simulations is the rejection-free 'residence-time' procedure, also referred to as the BKL or 'n-fold way' algorithm based on a publication in 1975 by Bortz, Kalos and Lebowitz [60]. Here, the use of only two random numbers is required independent of the amount of neighboring molecules. The first is used to determine which pathway is chosen to escape from state i. To this end all M pathways are identified with their rate constants k_{ij}. These are added up to give the total escape rate k_{tot}. A path with a low rate thus comprises only a small part of the total rate. This is implemented as an array of partial sums, where the array elements $s(j)$ represent the sums up to and including the rate k_{ij}, i.e.:

$$s(j) = \sum_{q}^{j} k_{iq} \qquad (3.33)$$

A random number r distributed on $(0, 1)$ is drawn and multiplied by k_{tot}. The array is then searched from the beginning until $s(j) > rk_{tot}$, which is identical to choosing pathway j for escape from state i. The second random number is required to advance the system clock and is drawn from the exponential distribution for the rate constant k_{tot}. Note that the time advance has nothing to do with which event is chosen. The time to escape depends only on the total escape rate. In the limit of an infinite number of hops both schemes are equivalent.

3.4.3. Determination of the rate constants

In our case the kinetic Monte Carlo simulations are applied to simulate charge transport in the hopping regime. It is assumed that due to weak correlations and strong disorder in organic systems the charge is always localized on one molecule and charge transport is governed by hopping rates, which define how likely it is to hop from a given molecule to a neighboring one. Rare event sampling and hence a KMC treatment is therefore applicable. The rates used correspond to the Marcus rates, which are explained in detail in sec. 4.4.2. The (hopping) sites represent centers of mass of different molecules or conjugated segments that a charge can hop to. The total hopping rate away from molecule (state) i is determined by adding up the Marcus rates for all nearest neighbors thereof. The algorithm implemented in our simulation code is identical to the one described above, if charge transport is simulated using a single charge only. In case of multiple charges, states already occupied will not be available to other charges.

4. Models for charge transport

In the past the study of the influence of chemical structure and morphology on charge transport properties for organic compounds has been impossible for two main reasons. First, it was unknown how to compute the necessary simulation parameters such as reorganization energies and transfer integrals for organic molecules. This has been solved by the advances in quantum chemistry outlined in sec. 3.1 and 3.2. Second, to obtain transfer integrals between neighboring molecules based on a realistic morphology requires large-scale molecular dynamics simulations and thus also the evaluation of a large number of transfer integrals. To achieve this at an affordable cost has become possible due to the molecular orbital overlap method.

These recent advances allow us to apply two state of the art charge transport models - kinetic Monte Carlo based on Marcus rates and diffusion limited by thermal disorder using semiclassical dynamics - to realistic morphologies sampled by MD. The approaches will be compared for organic crystals while focusing on hopping transport for liquid crystalline compounds. This enables us to test their abilities and limits and to find structure-property relationships aiding the understanding of charge transport. In this chapter we first give an overview of the development of charge transport descriptions starting from band theory up to the models used in this work. They will be explained in detail including analysis and implementation, which was partially written in the course of this work. Finally, we also describe how to compute all required input parameters.

4.1. Band theory

Charge transport is by nature of the electron a quantum mechanical process requiring a description based on the Schrödinger equation. Hence the computation of charge transport properties is an attempt to describe the motion of one or multiple electrons in a material by means of solving the Schrödinger equation for electrons in a possibly time-dependent potential defined by the material of interest. Since a solution of the full equation is not feasible, approximations have to be made. Most charge transport theories and all treated here ignore interactions between charge carriers and their influence on an externally applied electric field. The remaining Hamiltonian contains terms representing electrons, phonons (and thus the potential due to molecular structure and morphology of the compound) and the coupling between them:

$$H = H_{el}^0 + H_{ph}^0 + V_{el} + V_{el-ph}^{local} + V_{el-ph}^{nonlocal} + V_{impurities} \tag{4.1}$$

Here H_{el}^0 and H_{ph}^0 are the Hamiltonians for non-interacting electrons and phonons. Each localized state interacts with its neighbors by an electronic coupling term V_{el} and phonons are dispersionless, i.e. localized on each site. Sites may represent either single molecules in case of crystals or conjugated segments in case of polymers. The local electron-phonon coupling V_{el-ph}^{local}

corresponds to the reduction in energy of a molecule due to the presence of a charge carrier. This results from a geometric deformation of the molecule as well as a possible rearrangement of the surrounding (solvent) molecules. The non-local coupling $V_{el-ph}^{nonlocal}$ describes the influence of center of mass vibrations and the orientation of neighboring molecules with respect to each other on the electronic coupling between them.

Since conductivity was initially studied in inorganic crystalline solids, the potential of the crystal was taken to be periodic due to strong covalent bonds between atoms ensuring long-range order. In this case the carrier wave function is described by the Bloch functions

$$\Phi_{n,\mathbf{k}}(\mathbf{x}) = e^{i\mathbf{kx}} u_{n,\mathbf{k}}(\mathbf{x}) \tag{4.2}$$

where $\mathbf{k}$ is the wave vector, n the discrete band index and $u_{n,\mathbf{k}}$ is a function with the same periodicity as the crystal lattice. For any given n the corresponding states are called the electronic band and for each band there is a connection between its wave vector and its energy called band dispersion. Terms referring to electron-phonon coupling in eqn. 4.1 are ignored, the carriers are completely delocalized and charge transport is treated by an effective mass approach [61, 62]. In a one-dimensional array of molecules with only a single state, carriers with wave vector $\mathbf{k}$ are scattered to a new state $\mathbf{k}'$ by impurities or phonons. The time between collisions (mean relaxation time of the band states) J is taken to be constant and the carrier is assigned an effective mass of $\frac{1}{m_{eff}} = \frac{1}{\hbar^2}\frac{d^2\varepsilon}{dk^2}$. The mobility under such assumptions is given by

$$\mu = eJ/m_{eff} \tag{4.3}$$

For more general three-dimensional systems with multiple states per molecule, the mobility tensor is given by $\mu = eJM_{eff}^{-1}$, where the elements of the inverse mass tensor are given by

$$(M_{eff}^{-1})_{\mu\nu} = -\frac{1}{\hbar}\left(\frac{\partial^2 E(\mathbf{k})}{\partial k_\mu \partial k_\nu}\right)_{\mathbf{k}_0} \tag{4.4}$$

Here μ and ν represent one of the Cartesian coordinates x, y, z and $\mathbf{k}_0$ is the wave vector at which the lowest unoccupied (highest occupied) band energy of the electron (hole) is at its minimum (maximum). Despite the fact that organic crystals are hardly as ordered as their inorganic counterparts, band structure computations have been applied to organic crystals in an attempt to link the effective mass tensor with experimental mobilities [63, 64, 65] and give insights at low temperatures, but fail at room temperature. One result of the band transport model that holds for most organic crystals even at higher temperatures is that the temperature dependence of the mobility follows a power law:

$$\mu \propto T^{-\alpha} \tag{4.5}$$

Nevertheless, electron-phonon interactions cannot be neglected in charge transport calculations for organic materials and thus two different approaches to treat them were developed: polaron and disorder models. In the former, disorder is taken to be weak (as in crystals), so the wave function of the electron will still be spread out over a number of molecules requiring the solution of the underlying Hamiltonian. In the latter case, disorder is assumed to be strong or the coupling between the molecules weak (as in liquid crystals or polymers), so that the electron is localized on a single molecule and charge transport is governed by hopping rates between adjacent molecules. The corresponding rate equation may be solved numerically or using kinetic Monte Carlo simulations (see sec. 3.4).

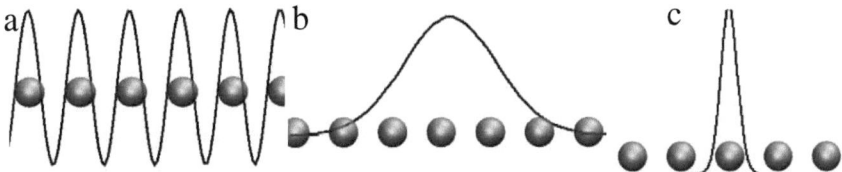

Fig. 4.1.: Equilibrium electronic wave functions in the band-like, fully delocalized, perfectly ordered case (a), the weakly disordered, polaronic case (c) and the fully localized, highly disordered case (c) illustrated on a one-dimensional array of molecules.

4.2. Polaron models

When the electron-phonon interaction can no longer be neglected, this means that the charge carrier deforms the surrounding medium when it travels. The quasi-particle formed by combining the charge carrier and the deformation it carries with it is termed polaron. An illustration of the different behavior of charge carriers in the limits of band-like, polaronic and hopping transport is given in figure 4.2.

4.2.1. Holstein models

The first to treat electron transport taking into account the local electron-phonon coupling in eqn. 4.1 was Holstein in 1959 [67, 68]. Charge transport is described along a one-dimensional array of diatomic molecules in the tight-binding approximation, i.e. the state of the system may be expressed as a linear superposition

$$\Psi(\mathbf{r}, x_1, \cdots, x_n) = \sum a_n(x_1, \cdots, x_n) \phi(\mathbf{r} - n\mathbf{a}, x_n) \qquad (4.6)$$

of molecular electron wave functions, $\phi(\mathbf{r}-n\mathbf{a}, x_n)$, each localized on a particular (nth) molecular site, with **a** being the unit lattice vector, and depending on the internuclear coordinate x_n of that site. Treating the potential energy of an individual molecule as parabolic, i.e. assuming harmonic vibrations, the Schrödinger equation of the system may be written as:

$$i\hbar \frac{\partial a_n(x_1, \cdots, x_n)}{\partial t} = \sum_{m=1}^{N} \left[-\frac{\hbar^2}{2M} \frac{\partial^2}{\partial x_m^2} + \frac{1}{2} M \omega_0^2 x_m^2 + A x_n \right] a_n - J[a_{n+1} + a_{n-1}] \qquad (4.7)$$

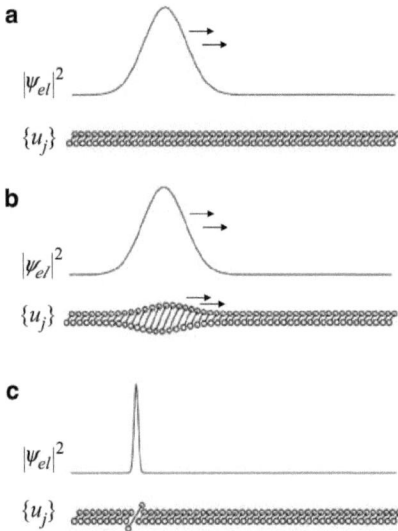

Fig. 4.2.: Illustration of charge carrier propagation in a one-dimensional array of diatomic molecules for different regimes, where deformations are indicated by modulations of the bond length u_j of the molecules. In pure band transport (a) the crystal lattice is unaffected by the carrier motion. In polaronic transport (b) the carrier travels along with a lattice deformation and is thus 'heavier' and slower. In the case of pure hopping (c) the lattice deformation due to the presence of the charge is so strong that the charge is completely localized on one molecule and hops between different molecules with a given rate. The transition between (a) and (b) is studied by large, the one between (b) and (c) by small polaron theories. Taken from [66].

Here the sum over m in the square brackets represents the lattice Hamiltonian, i.e. the vibrational potential and kinetic energy of a single molecule, where M is the reduced mass and ω_0 is the vibrational frequency of the individual molecules. $-Ax_n$ corresponds to the electron-lattice interaction. The deformation of a molecule occupied by a charge carrier is treated as a linear function of its vibration coordinate. The term proportional to the transfer integral J links the electronic wave function with molecular motion. Note that J is assumed to be identical between all sites, i.e. there is no off-diagonal disorder (non-local electron-phonon coupling). An important prediction of this theory is that the transfer integral is reduced by the electron-phonon interaction, in mathematical terms:

$$J(T) = J \exp\left\{-\frac{\Lambda}{\hbar\omega_0} \coth\left(\frac{\hbar\omega_0}{2k_BT}\right)\right\}, \qquad (4.8)$$

where Λ is the reorganization energy (sec. 4.5.2). It predicts a transition from band-like to hopping transport at higher temperatures. The mobility initially decreases with temperature due to band-width narrowing but increases again later due to phonon-assisted hopping. Based on this approach using the lowest order approximation for the relaxation time τ the mobility may be computed as in eqn. (53) of [24]:

$$\mu = \frac{ea^2\omega_0}{k_BT} \left[\frac{\Lambda \operatorname{csch}\left(\frac{\hbar\omega_0}{2k_BT}\right)}{\pi\hbar\omega_0}\right]^{1/2} \exp\left\{-\frac{2\Lambda}{\hbar\omega_0} \operatorname{csch}\left(\frac{\hbar\omega_0}{2k_BT}\right)\right\} \qquad (4.9)$$

with the hyperbolic cosecant $\operatorname{csch} x = 1/(\sinh x) = 2/(e^x - e^{-x})$. Due to the assumption of a narrow electronic band and a temperature larger than the electronic band width, this expression does not depend on J and is only valid if $J(T) \ll k_BT$.

4.2.2. Holstein-Peierls models

The inclusion of off-diagonal disorder, i.e. the Peierls or nonlocal electron-phonon coupling term from eqn. 4.1, was first computed without quantum chemical knowledge of the coupling parameters in 1985 by Munn and Silbey [69, 70]. Hannewald and Bobbert were able to give an analytical formula for the mobility in thermal equilibrium based on the small polaron picture by evaluating the Kubo formula for electrical conductivity, i.e. evaluating the current-current correlation function treating the total current as the sum of a purely electronic and a phonon-assisted current [71]. They then applied the model to naphtalene [72] by evaluating all necessary parameters using DFT-LDA calculations combined with MD simulations and were able to reproduce the experimental temperature dependence of the hole mobility $\mu \propto T^{-2.5}$. Nonetheless, it has been found that thermal fluctuations in transfer integral values between nearest neighbors in organic crystals are of the same order as their average value [73, 74]. Thus the delocalized nature of the carrier assumed in these approaches is no longer preserved due to localization produced by these fluctuations.

4.3. Diffusion Limited by Thermal Disorder

To include the strong nonlocal electron-phonon coupling, i.e. the large thermal fluctuations of the transfer integral J, a model must be used that does not rely on delocalization of the carrier. Due to the strong electronic coupling in crystals, a treatment by hopping models is also inappropriate. It is thus necessary to numerically evaluate the full Hamiltonian given in eqn. 4.1. The model Hamiltonian may be expressed as follows [66]:

$$H^0_{el} = \sum_j \epsilon_j a^*_j a_j \qquad (4.10)$$

$$H^0_{ph} = \sum_{kl} \hbar \omega_{kl} \left(b^*_{kl} b_{kl} + \frac{1}{2} \right)$$

$$V_{el} = \sum_{ij} J_{ij} a^*_i a_j$$

$$V^{local}_{el-ph} = \sum_{kl} \sum_j \hbar \omega_{kl} g_{jj,ql} (b^*_{kl} + b_{-kl}) a^*_j a_j$$

$$V^{nonlocal}_{el-ph} = \sum_{kl} \sum_{i \neq j} \hbar \omega_{kl} g_{ij,ql} (b^*_{kl} + b_{-kl}) a^*_i a_j$$

Here j labels the number of sites (molecules) in the system, a^*_j and a_j are the creation and annihilation operators of a charge carrier on site j, b^*_{kl} and b_{kl} are the creation and annihilation operators for a phonon with wave vector $\mathbf{k}$ in mode l with ω_{kl} being the corresponding frequency. J_{ij} is the transfer integral between neighboring molecules i and j. In the local electron-phonon coupling term $\frac{1}{2}\hbar\omega_{kl} g_{jj,ql}$ represents the energy reduction of a molecule in the presence of a charge carrier, which is equal to half the reorganization energy. Finally, the energy reduction in the nonlocal coupling corresponds to the modulation of the transfer integral induced by phonons. The reason this type of equation could not be tackled at an earlier point, even numerically, was the lack of knowledge of the parameters required to evaluate the Hamiltonian. The advances in computational chemistry outlined in sections 3.1 and 3.2 coupled with MD simulations (sec. 3.3) now allow the computation of all necessary parameters. The required computations are presented in detail in section 4.5.

4.3.1. Semi-classical Dynamics (SCD)

In this approach, diffusion limited by thermal disorder transport is treated by evaluating a one-dimensional model Hamiltonian using semi-classical dynamics identical to the model applied to rubrene by Alessandro Troisi in [75]. The Hamiltonian contains all terms present in eqn. 4.10 under the assumption that only the highest occupied (lowest unoccupied) molecular orbital $|j\rangle$ on each molecule j is relevant for the description of hole (electron) transport. The complete Hamiltonian then reads:

$$H = \sum_j \left(\epsilon + \sum_k \lambda^{(k)} u_j^{(k)} \right) |j\rangle \langle j| \qquad (4.11)$$

$$+ \sum_j \left(-J + \sum_k \alpha^{(k)} \left(u_{j+1}^{(k)} - u_j^{(k)} \right) \right) (|j\rangle \langle j+1| + |j+1\rangle \langle j|)$$

$$+ \sum_k \left(\sum_j \frac{1}{2} m^{(k)} \left(\dot{u}_j^{(k)} \right)^2 + \sum_j \frac{1}{2} m^{(k)} \left(\omega^{(k)} u_j^{(k)} \right)^2 \right)$$

Here the index j runs over all molecules in the one-dimensional array, k represents different nuclear vibrational modes with corresponding displacements $u_j^{(k)}$, effective mass $m^{(k)}$ and frequency $\omega^{(k)}$. Finally J is the transfer integral between two neighboring molecules when their nuclear modes are in equilibrium ($u_j^{(k)} = 0$).

The very first term is the sum of the energy ϵ of the electronic state relevant for the charge transport (HOMO for electron and LUMO for hole transport) and the Holstein electron-phonon coupling. The former may be obtained from quantum chemical calculations on isolated molecules. However, since it is the same for identical molecules, it is irrelevant for our charge transport simulations. The latter term $\lambda^{(k)} u_j^{(k)}$ takes into account the diagonal electron-phonon coupling, which is associated with the reorganization energy in small polaron theory. The second term takes into account only nearest-neighbor interactions. The transfer integral between two adjacent, identical molecules J is modified by the off-diagonal or Peierls electron-phonon coupling, where the displacement of two neighboring molecules with respect to each other $\left(u_{j+1}^{(k)} - u_j^{(k)} \right)$ modulates the transfer integral linearly with the coupling constant $\alpha^{(k)}$. The third term contains the classical kinetic and potential energies of the molecular vibrations, which are described as harmonic oscillators. In the simulations an initially localized carrier wave function spreads over the molecule array with time, yielding the mean-square displacement of the carrier and hence its mobility. The one-dimensional array of molecules along with the probability density at different times are shown in figure 4.3.

Obtaining the input parameters

In this model only two vibrational modes are taken into account: fast intra-molecular vibrations and slow inter-molecular vibrations with the corresponding frequencies $\omega^{(1)}$ and $\omega^{(2)}$. Treating the latter classically is certainly correct, while it is just as certainly incorrect in the former case, but since the intra-molecular vibrations only shift the absolute value of the mobility to lower values and not the temperature or field dependence, this is an acceptable approximation. A quick overview on how the input parameters are derived is given below, details on transfer integral and reorganization energy calculations are postponed to section 4.5.

- High frequency modes
  High frequency vibrations of molecules are obtained from QC frequency calculations. For phenyl rings and molecules composed thereof, they yield a number of modes with frequencies around $1400\,\mathrm{cm}^{-1}$ and very low effective masses compared to the entire mass of the molecule. The average frequency $\omega^{(1)} = 1400\,\mathrm{cm}^{-1}$ along with an effective mass of

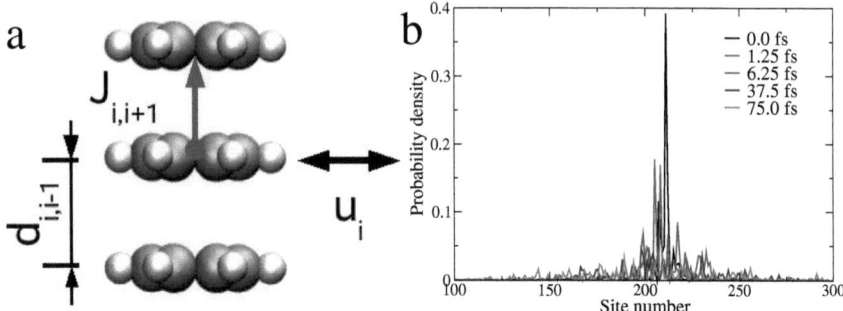

Fig. 4.3.: (a) In semi-classical dynamics the compound in question is modeled by a one-dimensional array of molecules. The average distance between the molecules $d_{i,i-1}$, the intersite coupling J_{ij} and the displacements of the molecules from their equilibrium position u_i are indicated. (b) Spreading of the initially localized wave function due to numerical evaluation of the Hamiltonian (eqn. 4.11).

6 amu corresponding to C-C bond vibrations in phenyls is chosen for the high frequency mode in our simulations. While the corresponding energy is high, it is almost independent from the external electric field or temperature and thus only shifts curves describing mobility dependencies without influencing their shape.

- Low frequency modes
  Low frequency modes correspond to vibrations of the center of mass of the entire molecule. The effective mass in this case is therefore simply the mass of the entire molecule. The energy fluctuations can be measured by a combination of MD and QC. MD is performed to sample a reasonable amount of time of the relatively slow vibrations. QC is then used to compute the transfer integrals between molecules for each snapshot, taken at very short time intervals (20 fs). The result is used to compute the autocorrelation function $< J(0)J(t) >$, the Fourier transform of which gives the vibrational frequency spectrum. The main peak average thereof corresponds to the low frequency mode used in the simulation.

- Average intersite coupling J
  This is simply the average of the transfer integrals obtained by the combined MD and QC simulations mentioned above for equivalent pairs.

- Holstein coupling constants $\lambda^{(k)}$
  The Holstein coupling constant λ is associated with the reorganization energy Λ and the Huang-Rhys factors, all of which may be computed as described by Brédas' group [76, 77]. Neglecting the external reorganization energy and assuming that all reorganization energy is carried by one effective high frequency mode, as was initially proposed by Jortner [78], the Holstein coupling constant is given by

$$\lambda^{(1)} = \omega^{(1)} \sqrt{m^{(1)}\Lambda} \qquad (4.12)$$

Since the slow vibrational frequency is a center of mass motion and does not perturb the structure of the molecule itself, it does not contribute to the reorganization energy and thus the Holstein coupling constant for the slow frequency $\lambda^{(2)}$ is zero.

- Peierls coupling constants $\alpha^{(k)}$
  The off-diagonal or Peierls coupling constant α is zero for high-frequency modes, because small vibrations of the molecular structure have little influence on the transfer integral between neighboring molecules. For the average low frequency mode $\omega^{(2)}$, the Peierls coupling constant is chosen such that the resulting distribution of transfer integrals at a given temperature T has the same standard deviation σ_τ as the one obtained from MD and QC calculations. This is ensured by matching the Boltzmann distribution of oscillator deviations $e^{-m\omega^2 x^2/2k_B T}$ in SCD with the Gaussian distribution of transfer integrals $e^{-\alpha^2 x^2/2\sigma_\tau^2}$ from MD using the following expression:

$$\alpha^{(2)} = \frac{\sigma_\tau}{\sqrt{2k_B T/m^{(2)}(\omega^{(2)})^2}} \qquad (4.13)$$

The numerical algorithm

The basic steps to numerically solve eqn. 4.11 are:

i. Compute initial positions, velocities and wavefunction.
   $u_j(t), \dot{u}_j(t), \psi(t)$

ii. Compute classical accelerations
    $m\ddot{u}_j(t) = -ku_j(t) - \frac{\partial}{\partial u_j}\langle \psi(t) | H^{el} | \psi(t) \rangle$

iii. Update wavefunction
     $\psi(t + \Delta t) = \psi(t) - iH^{el}\psi(t)\Delta t$

iv. Update position via Verlet algorithm
    $u_j(t + \Delta t) = 2u_j(t) - u_j(t - \Delta t) + \ddot{u}_j(t)(\Delta t)^2$

The positions and velocities of the harmonic oscillators at a given temperature follow a Boltzmann distribution $P\left(u_i^{(k)}\right) \propto \exp\left\{-k_i^{(k)} u_i^{(k)}/2k_B T\right\}$ with the standard deviation $\sigma = k_B T/k_i^{(k)}$, where k_i is the spring constant of the oscillator, $k_i = m_i \omega_i^2$. So both are drawn from a Boltzmann distribution in the initial step. Based on these values all terms of the Hamiltonian matrix can be calculated. The molecules only interact with their nearest neighbors in this model, i.e. the i-th molecule feels only the presence of the molecules $i - 1$ and $i + 1$, because the definition and calculation of transfer integrals is only sensible between nearest neighbor molecules. Thus the resulting matrix is symmetric, real and almost tridiagonal. Almost, because there are two additional entries at (1,N) and (N,1) to take care of the periodic boundary conditions. This Hamiltonian satisfies $H\Psi = E\Psi$ and thus the ground state energy is given by the lowest eigenvalue of the matrix. The corresponding eigenvectors are the values of the initial wave function coefficients. A single run is performed using the eigenvector corresponding to the lowest and thus groundstate eigenvalue E_0. For multiple runs, higher eigenvalues are chosen until $E_i - E_0$ significantly exceeds the thermal energy making it a negligibly unlikely starting state, because electron energy is Boltzmann distributed. In the main simulation loop the classical acceleration according to Newtonian mechanics is calculated first. For simple harmonic oscillators

4. Models for charge transport

$\ddot{u}_i = -k^{(k)} u_i^{(k)}(t)/m_i$, where the $u_i^{(k)}(t)$ are simply the current positions of the oscillators. The classical contribution to the potential energy is calculated via $V_{class} = 0.5 \cdot ku^2$. Next, the quantum mechanical contribution to the acceleration is computed and the corresponding energies $V_{qm} = \langle \Psi | H | \Psi \rangle$ are added to the classical potential energy.

$$a_{qm} = \frac{\partial}{\partial u_j^{(k)}} \langle \Psi(t) | H^{el} | \Psi(t) \rangle = \lambda^{(k)} C_j^* C_j + \alpha^{(k)} (-C_j^* C_{j+1} - C_{j+1}^* C_j + C_{j-1}^* C_j + C_j^* C_{j-1})$$

Thereafter, the time evolution of the coefficients is calculated. This is done using the quantum mechanical time evolution operator, i.e. $\dot{C} = -iHC$ and the product rule for the second derivative $\ddot{C} = -i(H\dot{C} + \dot{H}C)$. The new coefficients are then approximated by their second order Taylor expansion:

$$C(t + \Delta t) = C(t) + \dot{C}dt + \frac{1}{2}\ddot{C}dt^2$$

Following the coefficients, all the nuclear coordinates are updated. Starting with the positions $R(t + dt) = 2R(t) - R(t - dt) + a(t)dt^2$ and followed by the velocities $V(t + dt) = (R(t) - R(t - dt))/dt + \frac{1}{2}A(t)dt$, both required for the final step of calculating the new Hamiltonian and its time derivative. The Hamiltonian is calculated without explicit use of the wave function coefficients. Its time derivative is given by:

$$\dot{H} = \sum_j \left(\sum_n \lambda^{(k)} \dot{u}_n^{(k)} \right) |j\rangle\langle j| + \sum_k \sum_j \alpha^{(k)} (\dot{u}_{j+1}^{(k)} - \dot{u}_j^{(k)})(|j\rangle\langle j+1| + |j+1\rangle\langle j|)$$

Note that the wave function coefficients are not explicitely used to calculate the matrix elements, but to clarify which matrix elements are calculated. The algorithm is extremely efficient, because other than for the very first initialization of the coefficients no matrix diagonalization routines are needed.

Calculating the mobility

To obtain the mobility of the charge carrier based on the numerical simulations, the diffusion of the carrier is monitored quantitatively from:

$$R_n^2(t) = \langle \Psi_n(t) | r^2 | \Psi_n(t) \rangle - \langle \Psi_n(t) | r | \Psi_n(t) \rangle^2 \tag{4.14}$$

The resulting random mean square displacements for each initial wave function are Boltzmann-averaged and a fit to the resulting curve is made using the function $\frac{ax^2}{x+b} + c$ to incorporate the initial quadratic ballistic regime giving way to the linear diffusive regime at large times. The diffusion coefficient D is then calculated from $D = \lim_{t \to \infty} <R^2(t)>/2t$, i.e. a is twice the diffusion coefficient. The mobility μ may now be calculated using the Einstein formula:

$$\mu = \frac{eD}{k_B T} \tag{4.15}$$

Predictions of the model

SCD qualitatively shows the correct decrease of mobility with temperature $\mu \propto T^{-2.1}$ for rubrene [75] in accordance with band-like theories. It also shows a quadratic increase of mobility with average coupling $\mu \propto J^2$ when all other parameters remain the same, as is the case for hopping models. The main disadvantage of the current one-dimensional implementation of the model is that an increase in dynamic disorder (α) always leads to a decrease in mobility. The two latter dependencies are shown in fig. 4.4.

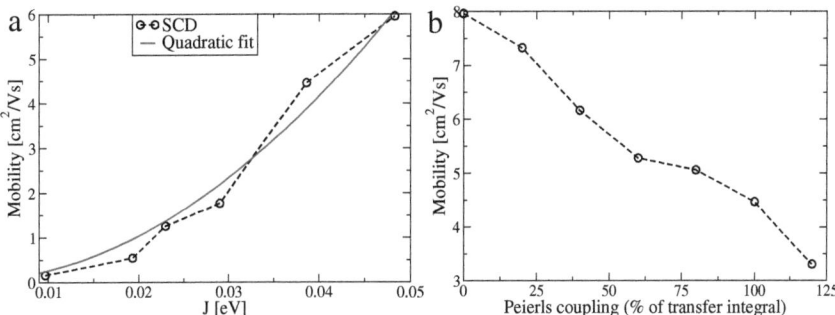

Fig. 4.4.: Dependence of mobility on transfer integral and Peierls coupling constant according to SCD. Scattering in the data is due to lack of statistics.

4.4. Disorder models

Disorder models assume that the charge carrier and its corresponding wave function are completely localized on one molecule. These are then treated as hopping site. The charge transport is thus governed by hopping rates between neighboring molecules.

4.4.1. Miller-Abrahams rates

In 1960 Miller and Abrahams described phonon-induced electron hopping between donor sites in doped semi-conductors [79]. They computed the transition rate based on electronic wave functions taken from effective mass theory [61, 62] by using a deformation potential for electron-phonon interactions. In their theory each hop had to be accompanied by emission or absorption of a phonon to conserve energy. Nonetheless, it came to be used to describe non-radiative transfer in organic materials later on with an additional Boltzmann factor for jumps upward in energy. Since we are using Marcus rates to simulate hopping transport in this work, we will simply state what is referred to as the Miller-Abrahams rate between two adjacent molecules i and j in the context of organic compounds [80]:

$$\omega_{ij} = \omega_0 \exp\left\{-2\gamma a \frac{\Delta \mathbf{R}_{ij}}{a}\right\} \begin{cases} \exp\left\{-\frac{\epsilon_j - \epsilon_i}{k_B T}\right\} & \text{if } \epsilon_j > \epsilon_i \\ 1 & \text{if } \epsilon_j < \epsilon_i \end{cases} \quad (4.16)$$

4. Models for charge transport

Here ω_0 is an empirical prefactor, a is the average lattice distance, $\Delta \mathbf{R}_{ij}$ is the distance between the two molecules, $S = 2\gamma a$ the orbital overlap parameter and $\epsilon_j - \epsilon_j$ is the difference in orbital energy between molecule i and j.

4.4.2. Marcus rates

For historical reasons, we will discuss the approach initially taken by Marcus. Other possible derivations are mentioned later. In his original work, which dates back to 1956 [81, 82], Marcus was interested in studying the electron transfer in isotopic exchange reactions such as $Fe^{2+} + Fe^{*3+} \rightarrow Fe^{3+} + Fe^{*2+}$ or cross reactions such as $Fe^{2+} + Ce^{4+} \rightarrow Fe^{3+} + Ce^{3+}$. It was a publication by Libby, explaining why isotopic exchange reactions involving electron transfer between pairs of small cations (like Fe^{2+} or Ce^{3+}) in aqueous solution were relatively slow while the same reactions between large ions (such as $Fe(Cn)_6^{3-}$ or MnO_4^{2-}) were quite fast, that caught his attention. Libby reasoned that when an electron is transferred from one reacting ion to another, the two new ions are formed in the wrong environment of the solvent molecules, because the nuclei do not have enough time to rearrange during the rapid electron jump (Frank Condon principle). For example a Fe^{3+} ion would be formed in the environment corresponding to an Fe^{2+} ion. Since for larger reactants the change of the electric field in the vicinity of each ion upon electron transfer is smaller, the original solvent environment appears less foreign to the newly formed ion and hence the energetic barrier of the reaction is lower. The mistake in Libby's treatment was that such a formation of an ion in the wrong high-energy environment could only happen by a vertical transition, i.e. the absorption of light. The processes occurred in the dark, however. Marcus showed that by random fluctuations in the nuclear coordinates of the ions and in the positions of the solvent molecules a configuration could be reached to satisfy both energy conservation and the Frank-Condon principle. He proposed the following reaction sequence leading from the reactants A and B to the products:

$$A + B \rightarrow X^* \qquad (4.17)$$

$$X^* \rightarrow X \qquad (4.18)$$

$$X \rightarrow \text{products} \qquad (4.19)$$

The reactants will fluctuate away form their equilibrium position ($A + B$) to form an activated complex X^*, corresponding to the atomic configuration which is at a maximum of the free energy with respect to the possible configurations. From X^* electron transfer may occur to form X and finally lead to the equilibrium configuration of the products. The most probable intermediate state may then be found by minimizing the free energy of formation ΔF^* of X^* subject to the restriction that X and X^* must have the same energy to fulfill energy conservation. This was done assuming spherical ions and a continuous description of the surrounding solvent. The expression for the rate k_1 to form the activated complex from the reactants is then given by:

$$k_1 = Z \exp\{-\Delta F^*/k_B T\} \qquad (4.20)$$

where Z is the collision number in solution. After testing this description on experimental data [83, 84], Marcus expanded it to allow treatment of arbitrary molecules in the reaction [85] resulting in the expression for a transfer rate between two molecules i and j:

4.4 Disorder models

$$\omega_{ij} = A \exp\{-\Delta G^*/k_B T\} \quad (4.21)$$

where the Gibbs free energy is given by:

$$\Delta G^* = \frac{\Lambda}{4}\left[1 + \frac{\Delta G^0}{\Lambda}\right]^2 \quad (4.22)$$

The A in eqn. 4.21 depends on the nature of the electron transfer reaction, ΔG^0 is the standard free energy of the reaction and therefore zero for self-exchange reactions and Λ is the reorganization energy (treated in detail in sec. 4.5) composed of solvational and vibrational components. The theory underlying the Marcus rates is often visualized by two parabolic curves representing the high-dimensional free energy curves of the reactants and the products. Note that near and at the point of intersection the individual curves do not actually continue unperturbed due to electronic interactions. Instead they split leading to a lower energy surface and a higher energy surface as shown in fig. 4.5. Classical Marcus theory ignores this effect in its derivation and is hence only valid for low electronic couplings, where the transfer integral J is significantly smaller than the reorganization energy Λ, i.e $J \ll \Lambda$.

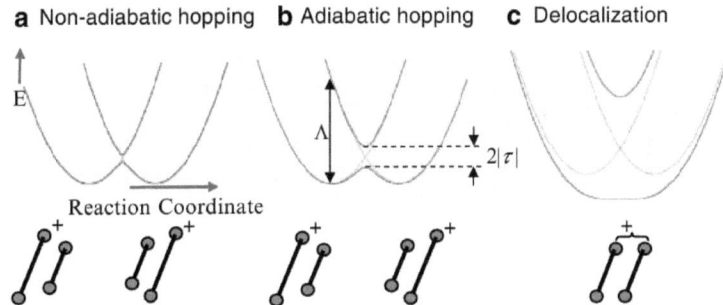

Fig. 4.5.: Effect of increased electronic coupling J on the free energy curves of two neighboring molecules. The two parabolas represent the diabatic energy for states where the charges are fully localized as assumed by Marcus theory. The thick lines are the adiabatic (Born-Oppenheimer) energy curves. Reorganization energy Λ and transfer integral J are indicated in the central panel. Going from left to right J is 0.03, 0.1 and 0.5 times the reorganization energy. Clearly, Marcus theory may only be applied for small J/Λ. Taken from [66].

An alternative approach to derive the Marcus rates is by utilizing the fact that the electronic wave function is localized on a single molecule and computing the rate of hopping between two molecules based on Fermi's golden rule [86], i.e. first order perturbation theory ($H = H_0 + V$):

$$\omega_{ij} = \frac{2\pi}{\hbar} \left\langle \Psi_i | V | \Psi_f \right\rangle^2 + \rho(E_f) \quad (4.23)$$

Here V is the perturbation of the two-molecule Hamiltonian due to electronic interactions, i and j are the initial and final electronic states and thus $\left\langle \Psi_i | V | \Psi_f \right\rangle^2$ is the electronic coupling matrix

4. Models for charge transport

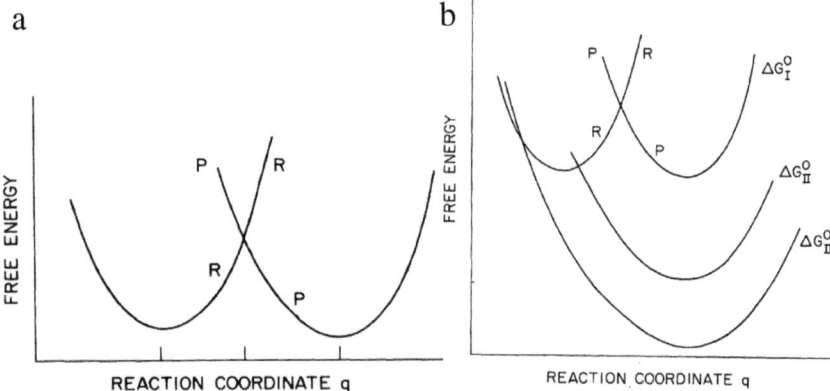

Fig. 4.6.: (a) Free energy of the reactants plus environment (R) and free energy of products plus environment (P) plotted versus a reaction coordinate q. The three vertical lines represent, from left to right, the equilibrium state of the reactants, the transition state and the equilibrium state of the products. (b) Plot of the free energy G versus the reaction coordinate q for reactants (R) and products (P) for three different values of ΔG^0 illustrating the inverted region effect, i.e. showing an increase in G for low ΔG^0 such as ΔG^0_{III} after an initial decrease of G upon lowering ΔG^0. Both figures taken from [82].

element or transfer integral J_{ij} between them (see section sec. 4.5). $\rho(E_f)$ is the continuous distribution of vibronically coupled final states, which is given by the Franck-Condon-weighted density of states under the assumption that all vibrational states are classical, i.e. in the high temperature regime where $\hbar\omega_i \ll k_B T$. Plugging this into eqn. 4.23 yields the most common Marcus rate expression for transport in organic materials used throughout this work:

$$\omega_{ij} = \frac{2\pi}{\hbar} |J_{ij}|^2 \sqrt{\frac{1}{4\pi\Lambda k_B T}} \exp\left\{-\frac{(\Delta G^0 + \Lambda)^2}{4\Lambda k_B T}\right\} \quad (4.24)$$

where J_{ij} is the transfer integral, Λ the reorganization energy, ΔG^0 the free energy difference between initial and final state and $k_B T$ the thermal energy. Note that this derivation does not require the assumption of an activated complex[78]. More sophisticated expressions taking into account the fact that not all vibrational modes are classical and valid for all temperature regimes were developed by Jortner and co-workers [87, 88]. Note that Marcus rates predict a decrease in rate and mobility at large ΔG^0, for example due to large applied electric fields, known as the inverted region effect. This effect is not present in the Miller-Abrahams rates, but simple to visualize in the free-energy curve picture. Successively making ΔG^0 more negative by lowering the products' G curve initially decreases the free energy barrier ΔG^* given by the intersection of the free energy curves for reactants and products. As shown in figure 4.6 the barrier eventually vanishes at a given ΔG^0 and is then seen to increase again. This effect was first seen experimentally in 1984 [89] and is well-known today. We chose to focus on Marcus rates in this work for several reasons. First, the two molecules in question need not be the same and may also be at different distances and orientations with respect to each other. Second, electronic devices function at ambient conditions corresponding to high temperatures. Third, at very low temperatures or very high coupling, polaron models are certainly more appropriate than hopping models.

4.4.3. Gaussian disorder model (GDM)

The use of kinetic Monte Carlo simulations to study charge transport in disordered organic compounds was first proposed by Bässler [80]. Based on the experimental knowledge that absorption bands of disordered organic solids are usually of Gaussian shape [90] molecules or conjugated segments of a polymer chain are treated as hopping sites and assigned site energies ϵ according to a Gaussian distribution:

$$\rho(\epsilon) = \frac{1}{\sqrt{2\pi\sigma^2}} \exp\left\{-\frac{\epsilon^2}{2\sigma^2}\right\} \quad (4.25)$$

Here σ refers to the standard deviation of the distribution chosen based on experimental data. This introduces diagonal disorder in the Hamiltonian, i.e. disorder due to electrostatics, polarization, geometrical shape or surrounding solvent. To take into account off-diagonal disorder, which may arise due to different distances or orientations of neighboring molecules, the intersite coupling S was also subjected to a distribution by splitting it into site specific contributions S_i and S_j, each taken from a Gaussian probability density. The latter was a somewhat arbitrary zeroth-order choice and also introduces correlations between jumps away from a given site. Miller Abrahams rates (eqn. 4.16) were used for the kinetic Monte Carlo simulations. Nonetheless, the model was capable of predicting a non-Arrhenius-type temperature dependence of the mobility μ:

$$\mu(T) = \mu_0 \exp\left\{-\left(\frac{2\sigma}{3k_BT}\right)^2\right\} \quad (4.26)$$

Upon taking into account an electric field tilting the density of states by an electrostatic potential, the Gaussian disorder model also predicts saturation of mobility at low fields along with an increase in a Poole-Frenkel fashion $\ln\mu \propto \sqrt{E}$ at higher fields. For vanishing disorder $\sigma = 0$ mobility was found to decrease with electric field, because the drift velocity saturates with applied field in a hopping system in which the field does not affect the intersite jump rates. Another prediction of the model was an increase in mobility upon an increase in off-diagonal disorder. This is explained by an overcompensation of the low rates by creation of high rate pathways and by the fact that a Gaussian distribution of the coupling has a non-Gaussian effect on the rates. It also showed an increase of mobility at low fields, again due to opening up of pathways with higher coupling, where some steps go against the field and are hence blocked at high fields. In the original model, dependence on charge carrier density was not taken into account. Due to the success of the model in fitting experimental data, improvements have constantly been added. In [91] six different variations of the Gaussian disorder model are presented, incorporating more sophisticated treatment of the off-diagonal disorder and taking into account the charge carrier density. Nonetheless, they all rely on Miller Abrahams rates and a Gaussian distribution of energies, both requiring parameters to be fitted to experimental values. Thus the models grant a basic understanding of charge transport, but are unable to make precise predictions on how to improve specific chemical compounds or morphologies.

4.4.4. Kinetic Monte Carlo with Marcus rates

To gain an understanding of the coupling of chemical structure and morphlogy with charge transport properties it is very important that the parameters required for kinetic Monte Carlo are based solely upon the underlying physical system. For this purpose extensive MD simulations must be coupled to the quantum chemical evaluations of the transfer integrals as was first done by Kirkpatrick et al. [92, 93, 94]. The morphology of ordered compounds must be known from x-ray experiments as starting point for sampling the system using MD. The snapshots of the equilibrated system are analyzed in the following steps:

  i. The molecules from the MD snapshot are mapped to coarse-grained beads consisting of their centers of mass and three orientation vectors.

  ii. Nearest neighbors are found using the simple search algorithm for triclinic boundary conditions [46].

  iii. Transfer integrals are calculated for all neighboring pairs using the molecular orbital overlap method [2] (sec. 4.5.3).

  iv. Hopping rates are computed between pairs based on their reorganization energy, free energy difference, transfer integral and a given temperature using Marcus theory.

  v. A KMC graph is built using the centers of mass of the molecules as vertices and the hopping rates as edges.

  vi. The mobility is calculated by time-of-flight, velocity averaging or diffusion analysis.

An illustration of the mapping procedure is displayed in figure 4.7. The mapping is done using the program package VOTCA [95], while the transfer integrals and kinetic Monte Carlo simulations are performed using a yet unpublished extension of the code written by Victor Rühle, James Kirkpatrick and myself. Below, we present the three different methods, which were developed to compute the mobility from KMC simulations.

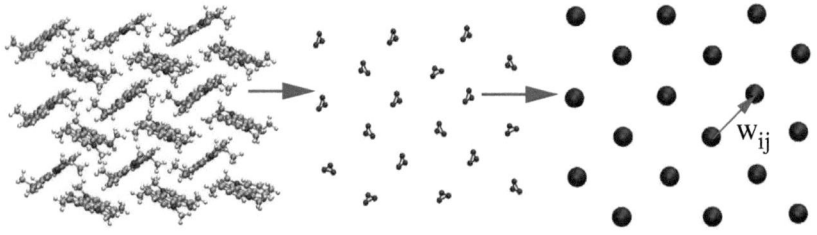

Fig. 4.7.: The two steps taken in the mapping from an MD snapshot to a KMC graph. First, molecules are mapped to coarse-grained beads and the hopping rates are computed for nearest neighbors. Second, a KMC graph is created based upon these rates and the centers of mass of the beads.

Time of Flight

Time of flight analysis was designed to yield transient currents similar to those obtained from actual time of flight experiments. To this end a supercell is generated from the initial MD simulation box to increase the transit time of the charges. An external electric field is applied in a given direction and charges are then generated in the first and collected in the last part of the supercell with respect to the field. Periodic boundary conditions are removed in direction of the field. During the kinetic Monte Carlo runs, the current is calculated and at the end plotted against simulation time. From the resulting curve, the transit time t_T may be obtained as the intersection of an extraplation of the plateau region and the tangent to the tail of the transient in a double-logarithmic plot (see sec. 2.2.1). Knowing the dimension of the supercell in field direction d and the applied electric field E yields the mobility:

$$\mu_{TOF} = \frac{d}{t_T E} \qquad (4.27)$$

Velocity Averaging

Both velocity averaging and diffusion analysis do not make use of the supercell and allow periodic boundary conditions everywhere. The difference between the approaches is schematically illustrated in fig. 4.8. In velocity averaging, the charge is allowed to travel for a given time and its velocity v is averaged throughout the simulation. It is then projected on the applied electric field, which allows calculation of the mobility via

$$\mu_{vel} = \frac{v}{E}, \qquad (4.28)$$

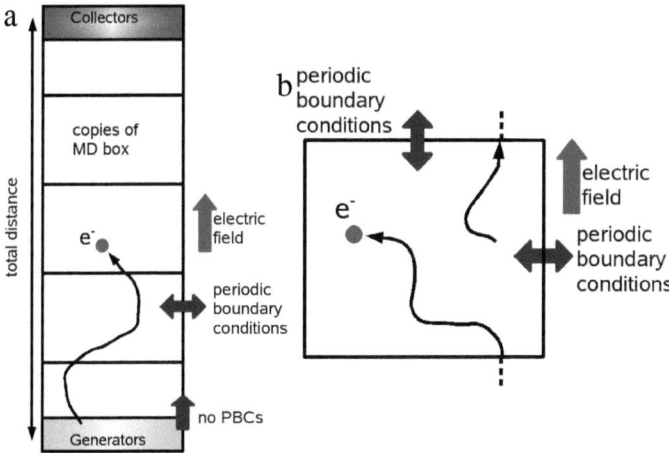

Fig. 4.8.: Schematic for KMC runs in case of time of flight (a) versus velocity averaging (b) analysis.

Diffusion

In the case of diffusion the external electric field is set to zero and the charge is allowed to explore the system for a given time t_{max}. At certain time intervals Δt the current position of the charge is recorded. The resulting random mean square displacement (RMSD) is calculated for each time interval Δt starting from the time between print outs dt up to the total length of the diffusion run t_{max}. Clearly, the statistics are far better for dt than for t_{max}, since for the latter only one interval exists, while for the former there are as many intervals as there are time intervals in the simulation. To improve the averaging for large time intervals, the RMSD analysis is performed for a hundred different starting positions. The final diffusion coefficient is calculated based on a regression over the first half of the Boltzmann average of all resulting curves. While an initial ballistic regime is present, it is too small to influence the validity of the regression. A typical diffusion pathway along with the fit to obtain the diffusion coefficient is shown in fig. 4.9. The mobility is computed based on the Einstein relation as for semi-classical dynamics (sec. 4.3.1). In general, the diffusion coefficient is a second rank tensor $D_{ij} = \lim_{t \to 0} \frac{<\Delta x_i \Delta x_j>}{2dt}$ with its highest eigenvalue and the corresponding eigenvector giving the highest mobility and the main direction for charge carrier diffusion.

$$\mu_{diff} = \frac{eD}{k_B T}, \tag{4.29}$$

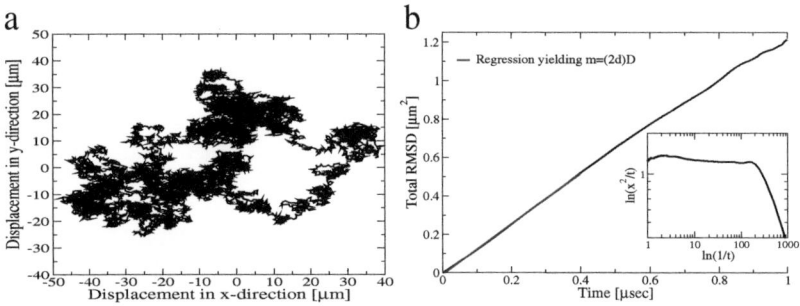

Fig. 4.9.: (a) A representative KMC diffusion pathway along with (b) the fit to the RMSD versus time to obtain the diffusion coefficient. The inset shows $<x^2>/t$ versus $1/t$ on double-logarithmic scale. The plateau represents the diffusive regime, at large times fluctuations occur due to statistical error, at short times the ballistic regime shows.

Predictions of the model

To ensure the correctness of the code we ran KMC simulations using a simple test system, where all molecules are equally spaced with every molecule having six neighbors, four in the same plane, one in direction of the field and one against the field. Transfer integrals between all neighbors are also identical. The temperature dependence of the mobility is obviously the same as for Marcus theory and electric field dependence shows the inverted field effect. The mobility saturates at low fields as expected, but simulations must be run significantly longer to remain accurate, because the difference between forward and backward hopping rates diminishes substantially with decreasing field. Thus longer runs must be performed to see significantly more forward than backward hops. When transfer integrals are taken from a Gaussian distribution with a standard deviation of σ to introduce off-diagonal disorder, the influence is seen to be low and leads to an initial increase in mobility followed by a decrease at higher disorder. Dependence of the mobility on electric field and off-diagonal disorder are shown in fig. 4.10.

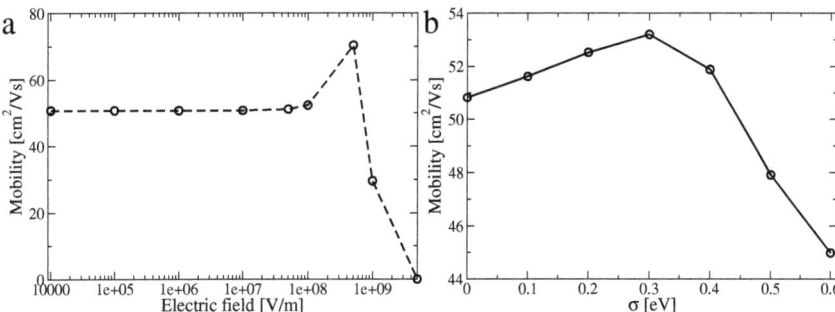

Fig. 4.10.: (a) Dependence of the mobility on electric field and (b) off-diagonal disorder implemented by a standard deviation σ of the transfer integral. The average transfer integral was taken to be 0.608 eV with a fixed molecule distance of 0.35 nm. Electrostatic disorder was not taken into account.

4.5. Charge transport parameters

One of the main challenges in describing charge transport for specific molecules in their different morphologies is to obtain the correct input parameters. MD simulations may be used to give an accurate picture of the morphology at a certain temperature, while Quantum Chemistry can be used to calculate all other relevant parameters. Since the importance of the correctness of these parameters cannot be overemphasized, we devote this section to the description of their computation.

4.5.1. Site energy

The site energy refers to the energy of the HOMO or LUMO level of the molecule, depending on whether one is interested in hole or electron transport. This is due to the fact that when an electron hops onto a fully occupied molecule it is most likely to enter the lowest unoccupied molecular orbital. Analogously, an electron is most likely to leave a molecule from the highest occupied molecular orbital and not from one of the inner shells. As is explained in sec. A.1.6 on restricted Hartree-Fock the n lowest eigenvalues of the Fock matrix correspond to the occupied ground-state orbitals of the molecule while higher eigenvalues correspond to excited states. This does not necessarily mean that the next higher orbital above the HOMO corresponds to the LUMO, since these orbitals are not truly taken into account in the SCF procedure and should be handled with care. However, more sophisticated methods yield reliable results and usually experimental data is also available for HOMO and LUMO energies.

4.5.2. Reorganization energy

The reorganization energy Λ consists of two parts: internal and external reorganization energy.

$$\Lambda = \Lambda_{int} + \Lambda_{ext} \qquad (4.30)$$

The external reorganization energy takes into account rearrangements of the surrounding medium, such as solvent molecules. Since there is no major rearrangement of molecules in crystals and none of the investigated systems contains solvent, the external reorganization energy is ignored throughout this work. The internal reorganization energy measures the energy cost of rearranging the geometries of the two molecules involved in the charge transfer upon transfer. Assume molecule i is initially charged while j is neutral with the corresponding energies E_{cc} and E_{nn}. Here the first index refers to the geometry and the second to the charge of the molecule. Once the charge hops from i to j, molecule i will be neutral but is still in charged configuration and molecule j will become charged but is still in neutral configuration with the respective energies E_{nc} and E_{cn}. The reorganization energy measures the cost to correct this, i.e. the energy difference between final and initial state:

$$\Lambda = (E_{nc} + E_{cn}) - (E_{nn} + E_{cc}) \qquad (4.31)$$

For a more detailed understanding, it is also possible to analyze which vibrational modes contribute to the reorganization [76, 86], but this is not of interest here. To calculate Λ, the charged and neutral geometries of a single molecule in vacuum are optimized using B3LYP with a 6-31G(d) or higher basis set. The two resulting geometries are then fixed and their wave functions are optimized in charged and neutral state using B3LYP/6-311G(d,p). This yields the four energies required above. Note that while the absolute energy values obtained from this calculation depend significantly on the basis set, the relative differences are hardly influenced by it.

4.5.3. Transfer integrals

The transfer integral is the most difficult to obtain quantity among the input parameters and much effort has been devoted to its computation [96]. It has been found to vary exponentially with distance for cofacial molecules, but also varies in a far more complex way with molecular orientation and thus fluctuates significantly with temperature. It alone can be used to explain the different mobilities found for identical molecules in different crystal phases as was done for the four different polymorphs of pentacene [97, 66]. The transfer integral may be defined as follows:

$$J_{ij} = \left\langle \Phi_i \left| H_{el} \right| \Phi_j \right\rangle \qquad (4.32)$$

where Φ is the multi-electron wave function with the indices indicating whether the charge is located on molecule i or molecule j and H_{el} is the electronic Hamiltonian of the system. When the multi-electron wave function is described by a Slater determinant (sec. A.1.5) and the frozen orbital approximation is invoked, i.e. only the HOMO (LUMO) level of the molecule influences the coupling for holes (electrons) and orbitals are not deformed in the presence of charges, we may evaluate the above equation for hole transport as:

$$J_{ij} = \langle \phi_{HOMO,i} | F | \phi_{HOMO,j} \rangle \qquad (4.33)$$

where F represents the Fock matrix (eqn. A.20) and $\phi_{HOMO,i}$ and $\phi_{HOMO,j}$ are the HOMOs of molecules i and j, respectively. When only two molecules are taken into account to represent the Hamiltonian of the system, which is justified if the molecules in the system are non-polar and long-range electrostatics can be neglected, the Fock matrix expressed as a linear combination of atomic orbitals (sec. A.1.3) takes the following form:

$$F = \begin{pmatrix} F_{ii} & J_{ij} \\ J_{ji} & F_{jj} \end{pmatrix} \qquad (4.34)$$

where F_{ii} and F_{jj} are localized on molecule i and j and J_{ij} represents the coupling between them. This is why differences in transfer integrals in a system are called off-diagonal disorder.

Energy-Splitting-in-Dimer

A straightforward approach to calculate the overlap integral is the energy-splitting-in-dimer (ESD) method [98, 24]. For two identical molecules in cofacial alignment the aforementioned Fock matrix simplifies significantly because now $F_{ii} = F_{jj} = F_0$ and $J_{ij} = J_{ji} = J$. Treating this problem as a two-state model, where the electron is either on molecule i or molecule j, we may calculate the respective energies by solving the secular equation:

$$\det \begin{vmatrix} F - E & J \\ J & F - E \end{vmatrix} = 0 \qquad (4.35)$$

This leads to $E_{1,2} = F \pm J$ and therefore allows computation of the transfer integral from the splitting of the HOMO (LUMO) energy $\Delta E = E_1 - E_2$:

$$J = \frac{\Delta E}{2} \qquad (4.36)$$

There is also an intuitive picture to this formula. The isolated molecules have identical HOMO levels. Upon approach the Pauli principle requires that two electrons may not exist in an identical state, thus the electronic levels split upon approach. The closer the molecules approach each other, the larger the energy level splitting and the transfer integral J become. The substantial drawback of this method is that it is invalid for different molecules or orientations strongly deviating from the cofacial one. Though techniques exist to correct for this [99], the method is hardly suitable for our intended computations.

Projective method

Another possible approach is to use the spectral theorem to project the molecular orbitals (MOs) of the pair onto a basis set defined by the individual molecules [2]. Knowing the eigenvalues of the MOs of the pair, the Fock matrix may then be reconstructed in the basis set of the individual molecules. J can then be read from the appropriate position in the new Fock matrix. The basis set of the individual molecules C_{loc} is written such that the first $N/2$ atomic orbitals are localized on molecule i and the second $N/2$ on molecule j. The same holds for the molecular orbital labels. The transformation matrix to convert from the basis set of the pair to the localized one is given by $C_{pair}^{loc} = C_{loc}^T C_{pair}$, where T stands for the transposed matrix. The Fock matrix in the localized basis set is thus given by: $F^{loc} = C_{pair}^{loc\,T} \epsilon_{pair} C_{pair}^{loc}$, where the eigenvalues ϵ_{pair} have been written in diagonal matrix form. The transfer integral between the HOMOs of molecules i and j, assuming the HOMO on molecule i is the nth orbital, is given by the Fock matrix element

$$J = F^{loc}_{i,i+\frac{N}{2}} \quad (4.37)$$

Since dimer calculations take a significant amount of time, this method is also unsuitable for computing the transfer integral between all neighboring pairs of an MD snapshot. It is, however, perfectly valid no matter what the molecular orientations may be.

Molecular orbital overlap (MOO) method

When examining the result from the projective method (eqn. 4.37), it becomes clear that only a calculation of the off-diagonal blocks of the Fock matrix is necessary to obtain the transfer integrals. As was shown in [2] all these off-diagonal elements will be of the form $F_{\mu\nu} = \bar{S}_{\mu\nu} \frac{\beta_i + \beta_j}{2}$, where $\bar{S}$ represents the overlap matrix of atomic orbital overlaps with differently weighted σ and π overlap, i and j label the two atomic centers that the μ and ν orbitals are centered on and β_i labels the bonding parameter of atom i. Instead of determining the Fock matrix, it thus suffices to compute $\bar{S}$, the weighted atomic orbital overlap. ZINDO (sec. A.1.7) is used for the overlap calculations and the necessary weighting. This can be done without requiring an SCF calculation on the pair of molecules and is therefore very fast, which is also the reason why it is the method of choice for this work. To check the validity of the method we compared it with ESD for two cofacial carbazole monomers at different distances. The agreement was found to be excellent and also shows the exponential decay of J with distance (see fig. 4.11).

4.5.4. Change in free energy

The change in free energy ΔG may be composed of the following terms:

- Site energy difference
  The difference between the HOMO (LUMO) levels of the two molecules i and j involved in the charge transfer. Taken to be zero for identical molecules in our simulations, while energy levels of identical molecules may also be chosen from a Gaussian distribution as in the GDM (sec. 4.4.3). Since the site energies are eigenvalues of the Fock matrix, differences in site energies are called diagonal disorder.

$$\Delta G = \epsilon_i - \epsilon_j \quad (4.38)$$

4.5 Charge transport parameters

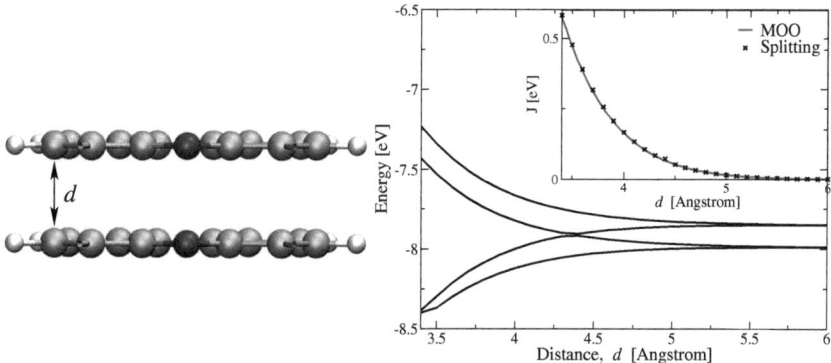

Fig. 4.11.: Splitting of the HOMO energy level as a function of separation between two carbazole monomers. The inset shows a comparison of the transfer integral values with results of the molecular overlap method. Fitting to a simple exponential decay $\exp(-\alpha d)$ yields $\alpha \approx 0.176 \, \text{nm}^{-1}$. Taken from [100].

- Electric field
  When an external electric field is applied, it will be favorable for the charge to travel in direction of the field. The further the better so the energetic contribution for a jump from molecule i to molecule j is

$$\Delta G = -e\mathbf{E}\mathbf{r}_{ij} \qquad (4.39)$$

- Electrostatic energy
  To compute the electrostatic site energy difference due to a pair of neighboring molecules i and j, the total electrostatic energy of an entire MD simulation box is calculated in case molecule i is neutral and in case it is charged. The difference in energy V_i between the two calculations is the electrostatic energy of molecule i. In mathematical terms the electrostatic energy between two molecules i and j can thus be expressed as $V(i^{(\alpha)}, j^{(\beta)})$, where i and j are the molecular indices and α and β describe whether the molecule is in charged (c) or neutral (n) state. This gives:

$$V_i = \sum_{j \neq i} V(i^{(c)}, j^{(n)}) - V(i^{(n)}, j^{(n)}) \qquad (4.40)$$

The electrostatic site energy difference contributing to ΔG_{ij} in the Marcus rate is thus given by $V_j - V_i$. The partial charges of the atoms required for above calculations are obtained by use of CHELPG (sec. 3.3.2) for neutral and charged configurations of the molecules optimized in vacuum.

- Polarization
  The influence of dipole and quadrupole interactions and the difference in their alignment prior and after the charge transfer is more tedious to calculate, but expected to have a low

contribution unless there are small and flexible solvent molecules or large dipoles in the system. Since neither is the case for the investigated compounds, this effect is ignored here. Both electrostatic energy and polarization may also be considered as part of the external reorganization energy, but recalculating it for each individual pair based on the MD snapshots is certainly the more accurate treatment.

5. Carbazole macrocycle

Two main problems in the construction of efficient organic solar cells shall be addressed in the following. Firstly, the excitons formed by the incident photons must be quickly separated into electrons and holes and secondly, the two charge carriers must be transported efficiently toward the electrodes (see sec. 2.1.1). In organic compounds, a light-generated exciton has a short lifetime and hence diffusion length before it recombines and cannot contribute to the photocurrent. It must therefore be able to quickly reach an energetically favorable acceptor molecule, where the electron transfers to the acceptor and the hole remains on the donor, forming a Coulomb-bound polaron pair. The polaron pair is then separated by the electric field arising from the equalization of the Fermi levels of anode and cathode. However, efficient charge percolation pathways must exist for both holes and electrons to allow them to reach the contacts. The construction of a π-conjugated macrocycle around a central π-system is able to address both issues simultaneously. The inner part may act as donor and the cycle as acceptor or vice versa. Hence we have a donor-acceptor system on the length scale of exciton diffusion, allowing efficient charge separation. If this ring-core structure additionally assembles into a columnar superstructure, it forms a coaxial cable, which gives rise to separate charge transport pathways. Thus efficient transport for both holes and electrons is possible due to the one-dimensional side-by-side percolation pathways along the columns. An alternative to coaxial cables are donor-bridge-acceptor systems treated in depth in refs. [101, 102].

A system with the potential to serve as a coaxial cable is the shape-persistent, π-conjugated macrocycle synthesized in Prof. Müllen's group [103]. It consists of polycarbazole, a material commonly used as a donor in organic solar cells and chromophores [104, 105], which has been built around a porphyrin template to form a carbazole macrocycle. In the bulk and after removal of the porphyrin template, the macrocycles form columns arranged on a hexagonal lattice, which allows for charge carrier transport in the π-stacking direction. It has only recently been shown that the cavity inside the conjugated carbazole macrocycles may also be filled by insertion of graphene molecules [106]. Carbazole itself is a cheap raw material, which may be obtained from coal-tar distillation [107]. Oligomers, polymers and small molecules containing carbazole are known to form relatively stable radical cations, exhibit high charge carrier mobilities and good thermal and photochemical stability [108]. Carbazole derivatives are widely used in the field of organic electronics as hole-conducting materials with a wide energy gap, e. g. carbazole-based polymers were employed to build dye solar cells [109] and carbazole-based small molecules were employed in blue organic light emitting diodes [110].

We have thus decided to study the morphology and charge transport properties of carbazole macrocycle columns arranged on a hexagonal lattice to reveal the influence of fluctuations in the columnar structures as well as in the macrocycles on the charge mobility. To this end, we use a combination of molecular dynamics simulations with kinetic Monte Carlo based on Marcus rates, as it has been shown in the past that this approach describes charge transport in columnar discotic systems well [92, 93, 94, 22]. Another interesting and theoretically challenging aspect

5. Carbazole macrocycle

is the treatment of the charge conjugation length along the macrocycle. In polymeric systems, the charge delocalizes on conjugated segments. Conjugation in polymers is broken mainly due to the torsional angle between neighboring monomers [111], hence we define different charge transfer units corresponding to the conjugated segments and use different rates for intra- and interchain transport. We chose to focus on the two extreme cases: first, full conjugation of the macrocycle, and second, conjugation broken between consecutive monomer units.

In the following, we will first describe how to compute the necessary force field parameters to describe carbazole macrocycles, which do not exist in the OPLS force field. We then give details on the molecular dynamics simulations in sec. 5.2, followed by the quantum chemical calculations to obtain the charge transport parameters in sec 5.3. Thereafter, we describe the results of the rate-based charge transport simulations and compare the two different conjugation lengths.

5.1. Force field parameters

The atomistic structure of a carbazole macrocycle repeat unit along with the atomic labels used to define the force field parameters are shown in fig. 5.1a. Despite the attention carbazole receives in the organic electronics community, the force field parameters to model carbazole are not readily available in the standard as well as the OPLS force field. In the following, we therefore exemplarily discuss in detail how an appropriate set of force field parameters can be determined. As a starting point we took the OPLS parameters for indole [112] due to the structural similarity with carbazole (see fig. 5.1b). The alkyl side chains were modeled using the OPLS united atom force field [113, 92]. This lead to the definition of the following eight atom types for the description of the carbazole repeat unit: the nitrogen (N_A), the carbons linking to it (C_N), the remaining carbons in the central five ring (C_B), the phenyl ring carbons with hydrogens attached (C_A), the corresponding hydrogens (H_A) and the CH_2 and CH_3 united atoms of the side chain (see fig. 5.1c). For polycarbazole and thus for the macrocycle the carbons linking two monomers (C8 & C9 in fig. 5.1a) were also assigned a unique type (C_C).

To describe polycarbazole macrocycles five sets of parameters are not available in the OPLS force field: the angle between core and side chain (C_N-N_A-CH_2), the dihedral potential describing the rotation of the side chain with respect to the core (C_N-N_A-CH_2-CH_2, angle Ψ in fig. 5.1c), the improper dihedral potential ensuring a ninety degree angle between core and side chain (N_A-CH_2-C_N-C_N), the dihedral potential describing the rotation around the central bond of a dimer, i.e. of two monomers with respect to each other (C_A-C_C-C_C-C_A, see Ψ in fig. 5.1d) and, finally, the improper dihedral potential ensuring planarity (C_C-C_C-C_A-C_A). The side chain atoms are labelled C14, C15, C16 and so forth. In the computation of the dihedral potential parameters between side chain and core, the side chain was shortened to four molecules, because the length of the side chain does not influence the dihedral potential parameters. Hence C17 became the last side chain atom of type CH_3. For the calculation of the dihedral potential parameters between two monomers, i.e. when a bond exists between C9 of the left and C8 of the right monomer, the entire side chain was replaced by a single hydrogen atom H1 linking to N1.

The calculation of the force field parameters was performed in accordance with the methods described in sec. 3.3.1. All calculations were performed using the *Gaussian* program [114].

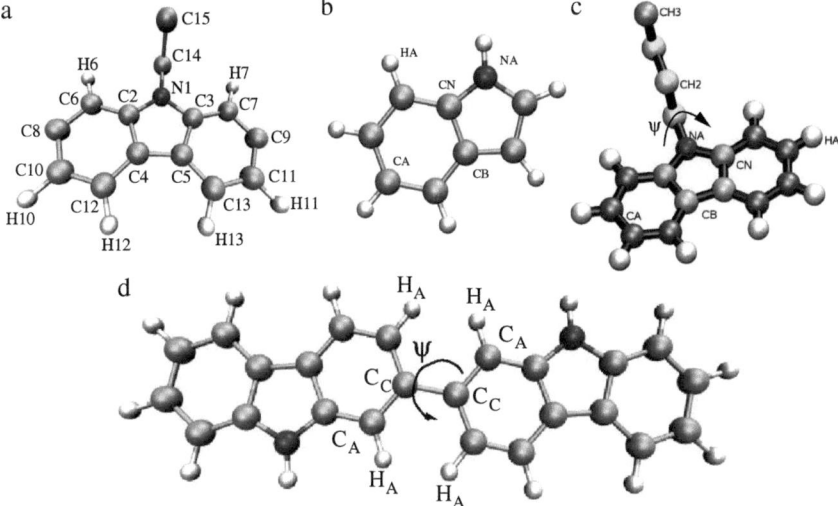

Fig. 5.1.: (a) Atom labels in a carbazole repeat unit (monomer). The bonds to other units in the macrocycle originate from C8 and C9. In the single unit there are two additional hydrogens (H8 and H9) instead. (b) Indole, from which most force field parameters were taken. (c) Carbazole single unit with an attached alkyne side chain and atom types used in the force field. The cores of both (b) and (c) are conjugated. (d) Carbazole dimer in optimized configuration, i.e. with the central dihedral angle Ψ (C_A-C_C-C_C-C_A) at 45°. Non-bonded interactions between hydrogens labelled by H_A were excluded in the force field computations.

Fitting of the parameters was based on structures optimized with B3LYP/6-311G(d,p). Self-consistent field energy calculations (see sec. A.1.4) were performed using the aug-cc-pVDZ basis set. The results of these calculations for the change in total potential energy of the system upon rotation of the side chain and dimer dihedral are shown in fig. 5.2. The potential of a single carbazole unit with respect to rotation of the side chain has its equilibrium configuration at a 90° angle to the core. Due to the importance of the dimer dihedral for the carbazole macrocycle structure and fluctuations, DFT using the PBE functional for both correlation and exchange energy and hybrid DFT/HF using B3LYP with two different basis sets were compared. They all reproduce the same minima and maxima and thus the same equilibrium configurations. We focus on results obtained with the B3LYP functional, which is a compromise between computationally expensive perturbation methods such as MP2 and DFT functionals less suitable for organic compounds such as PBE. For the potential of the dimer upon rotation of its central dihedral, there is a competition between electrostatic interactions and improvement of conjugation across the central bond. This leads to two minima at 45° and 135° of almost the same depth (see fig. 5.2b). When the nitrogens face in different directions, the molecules align in a linear configuration and repetition of multiple units leads to a polymer chain. When the nitrogens face in the same direction, however, the dimer makes a slight turn and repetition of twelve monomers forms a complete circle, i.e. the macrocycle. The height of the maxima in the potential energies due to side chain or dimer dihedral rotation are eight or four times the thermal energy at room temperature. Neither dihedral is thus expected to flip in the course of an MD simulation.

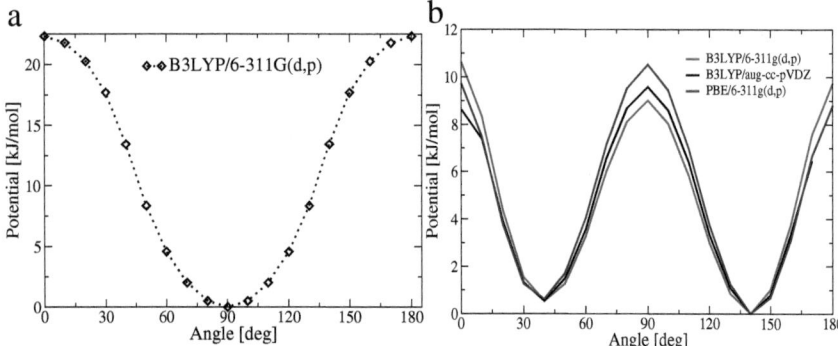

Fig. 5.2.: (a) Ab initio calculation results for the change in potential energy of a single carbazole unit upon rotation of the dihedral between core and side chain. (b) Comparison of potential energies due to the rotation of the dihedral in the center of a carbazole dimer calculated using PBE and B3LYP with two different basis sets. Dihedral energy calculations were performed in 10° intervals.

Since the QC methods calculate the change in energy of the entire system including contributions due to changes in bond lengths, angles, impropers, other dihedrals and long-range interactions, all these contributions must be subtracted from the QC results in order to isolate the potential of the dihedral in question. To achieve this, the energy of each configuration is calculated using the existing OPLS force field parameters with the parameters for the dihedral in question set to zero. We compared two different approaches to do this computation. The first uses the exact QC optimized geometries for each angle to compute the respective force field energies. The second takes these geometries as starting point for an energy minimization using the steepest descent method but fixing the dihedral in question. The advantage of the former method is that the geometries are closer to the physical reality. The disadvantage is that the actual rotation of the dihedral in the MD simulations will never occur along these geometries, but along those obtained by the latter method. Therefore to reproduce the torsional potential in MD simulations the steepest decent optimization method is more appropriate. All force-field calculations were done using the *Gromacs* program [46]. In accordance with the OPLS all atom force field, Lennard-Jones and electrostatic 1-4 interactions were scaled down by a factor of 0.5. When evaluating the force-field energies, the long range (1-4 and 1-5) interactions between carbons within the same benzene ring were excluded, because they share electrons and the interaction between them is of purely quantum mechanical nature and cannot be approximated by simple analytical terms. Additionally, for the single unit interactions between the second side chain atom (C16) and the two closest core hydrogens (H6, H7) were excluded, because at closest approach the classical description is no longer valid and yields too high values in energy. This does not mean that their interactions are no longer taken into account, but that they are included as part of the dihedral potential instead of as part of the long-range interactions. The resulting potential energies calculated using the force field for the geometries with and without relaxation using steepest decent are compared in fig. 5.3. The curve based on the exact QC geometries shows an unphysical kink at the maximum, because the configurations will never be sampled in the force field, hence the curve based on the relaxed geometries is chosen as a reference as well as for the dimer and all parameter calculations in the following chapters. For the dimer, the interactions of the four closest hydrogen atoms between the two carbazole units with each other were excluded, i.e. the interactions: H9 +H8, H9 +H10, H11 +H8 and H11 +H10, where the plus sign indicates that the second hydrogen of the pair belongs to the next carbazole monomer. The scanned angle and the excluded hydrogens are shown in fig. 5.1d.

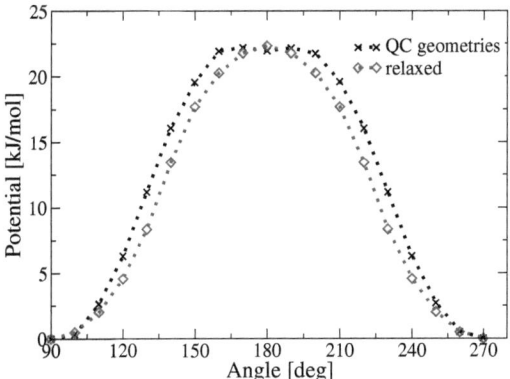

Fig. 5.3.: Change in potential energy of a single carbazole unit due to rotation of the side chain dihedral C_N-N_A-CH_2-CH_2 calculated based on the original B3LYP/6-311G(d,p) geometries as well as on those relaxed via steepest decent algorithm using the existing force-field parameters with the C_N-N_A-CH_2-CH_2 dihedral potential parameters set to zero.

The final parameters used in the force field are obtained by fitting to the difference between QC

5. Carbazole macrocycle

and force-field based energies. For the calculation of the latter the parameters of the dihedral in question were set to zero. The Ryckaert-Belleman (RB) function, which is a cosine expansion to fifth order, is used to fit the potential:

$$V_{RB}(\psi) = \sum_{n=0}^{5} V_n \cos^n \psi \quad (5.1)$$

Fig. 5.4 shows the resulting potentials for the side chain and dimer dihedral based on B3LYP QC computations and force field calculations using relaxed geometries. In case of the side chain dihedral, a standard fit does not represent the low energy regions well. Since they are the most important, because the system spends most of the time in these regions, fitting with a Boltzmann weighting factor of $\exp\{-E/(k_B T)\}$ was used, where $T = 300$ K or $k_B T = 2.4944$ kJ/mol, yielding significantly better agreement in the low energy regions. For the central dimer dihedral a standard fit was sufficient.

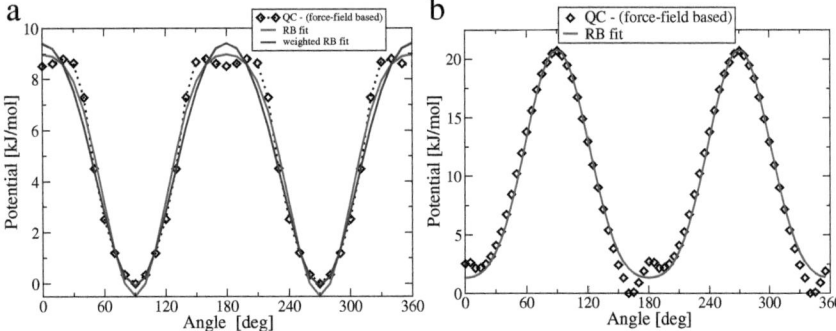

Fig. 5.4.: Fitting of Ryckaert-Belleman function to difference between B3LYP/aug-cc-pVDZ and force-field based energies for (a) the side chain and (b) the central dimer dihedral.

To validate the newly calculated force-field parameters, the force-field based energies are compared to the QC energies this time including the newly calculated dihedral parameters. The agreement shown in fig. 5.5 is found to be very satisfactory.

In addition to the dihedral potential parameters, improper and angle potential parameters were also calculated by a similar procedure. The fitting was performed using a harmonic potential $V(\phi) = \frac{1}{2} k_\phi \cdot (\phi - \phi_0)^2$. In case of the angular potential, the change in energy of other angles of the same type had to be taken into account. A complex multi-dimensional fitting procedure could be avoided, because scanning a particular angle only lead to small linear changes in other angles of the same type. For example to calculate the force-field parameters for the C_N-N_A-CH_2 angle, scanning of C2-N1-C14 lead to a small linear change in C3-N1-C14. Both angles must be described by the same force constant, thus the fitting function used was $V(\phi) = c + 0.5 \cdot k_\phi \cdot [(\phi - \phi_0)^2 + (3.9076 - 0.782 \cdot \phi - \phi_0)^2]$, where $3.9076 - 0.782 \cdot \phi$ describes the linear change in C3-N1-C14 due to rotation of ϕ, i.e. C2-N1-C14. Ignoring the second angle in the fitting would have lead to a far greater force constant and an unsatisfactory agreement between QC and force-field results, as is seen by comparing the red and the black curve in fig. 5.6.

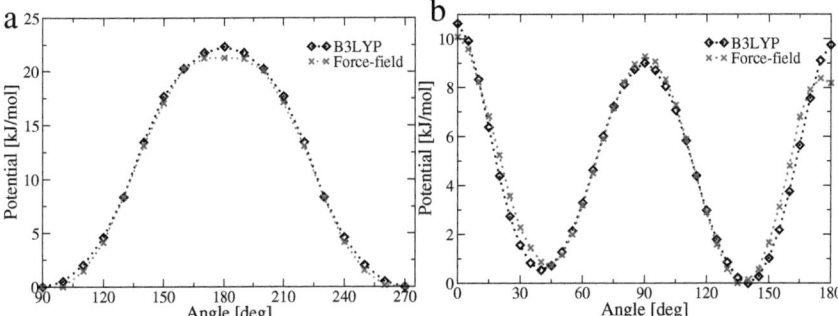

Fig. 5.5.: Comparison between force-field based energies using the new dimer dihedral potential and QC calculations based B3LYP/aug-cc-pVDZ for (a) the side chain and (b) the central dimer dihedral.

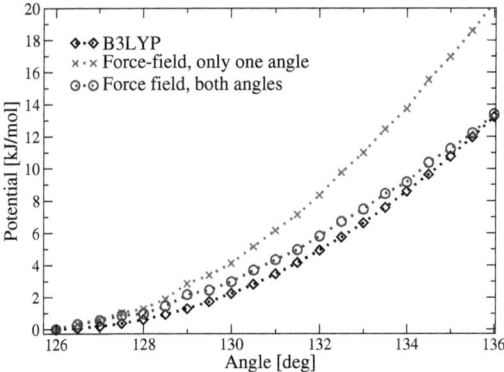

Fig. 5.6.: Comparison of force-field energies based on new angular potential with initial QC calculations using B3LYP/6-311G(d,p). Fitting procedures taking into account both angles of the same type and only the scanned angle are compared.

All fitting procedures required one and only one set of parameters to be set equal to zero, i.e. all other parameters had to be known. This is another reason why we chose to compute the dimer dihedral potential without a side chain attached to the monomers. Angular and improper parameters were initially taken from similar compounds present in the OPLS force field. The dihedral potential was then computed. Thereafter, angle and improper potential were computed and in a last self-consistent step the dihedral angle was computed again with the new angle and improper values. Results displayed in table 5.1 refer to the final dihedral computation. All equilibrium positions for bonds and angles in the force-field were checked against x-ray data and QC calculations. Deviations were found to be smaller than two percent in most cases, otherwise values were adjusted. It was also ensured that angular sums add up to the correct values in phenyl and five rings, i.e. to 720° or 540°, respectively. The final force field parameters are listed in tables 5.1 and 5.2.

dihedral	V_1	V_2	V_3	V_4	V_5
$C_N - N_A - CH_2 - CH_2$	0.0	0.735	0.0	8.120	0.0
$C_A - C_C - C_C - C_A$	-0.339	-34.40	-0.732	24.16	0.411
$CH_2 - CH_2$	17.45	1.14	-27.20	0.0	0.0

Tab. 5.1.: Coefficients of the Ryckaert Belleman function (eqn. 5.1) for the side chain and dimer dihedrals in kJ/mol. Note that they differ from the values published in [100], since at that point we did not do a steepest decent optimization prior to fitting. The CH_2-CH_2 dihedral is taken from previous HBC calculations [93].

angle	ϕ_0	k_ϕ
$C_N - N_A - CH_2$	126.4	153.21
$C_N - N_A - C_N$	111.6	585.76
$C_A - C_B - C_B$	134.9	711.28
$C_B - C_B - C_N$	108.8	711.28
$CH_2 - CH_2 - CH_2$	109.5	502.1

improper	ϕ_0	k_ϕ
$N_A - CH_2 - C_N - C_N$	0.0	114.0
$C_C - C_C - C_A - C_A$	0.0	761.7
$C_A - C_A - C_A - C_A$	0.0	167.4
$C_A - H_A - C_A - C_A$	0.0	167.4

Tab. 5.2.: Equilibrium angles ϕ_0 in degrees and force constants k_ϕ in kJ/mol for angles and impropers in the carbazole force field. Only the first angle and the first two impropers were computed bottom-up using the above approach. Side chain angles were taken from [93].

Having computed the bonded, short range interactions missing in the OPLS force field, treatment of the long range interactions remains. The computation of long range Coulomb interactions in the OPLS force field relies on the knowledge of the partial charges of each atom in the molecule. As outlined in sec. 3.3.1 convergence of the partial charges with respect to basis set was checked starting from 3-21G up to 6-311++G(3df,3pd). Charges were computed for a single molecule in vacuum for the force field computations and also for an entire macrocycle in vacuum for the final MD simulations, where the highest basis set used was 6-311G(2df,2pd). In the latter case, only the first united atom of the side chain was assigned charge (C14) while the other were considered neutral. In the simulated macrocycle, side chains with OH end groups also appear (see fig. 5.7), whose partial charges are taken into account. The convergence with respect to the basis set size for two chosen atoms for a repeat unit are given in tab. 5.3. The final choice of parameters corresponds to the value of the highest basis set, modified such that the symmetry of the partial charges exactly matches that of the molecule.

5.2 Molecular dynamics simulations

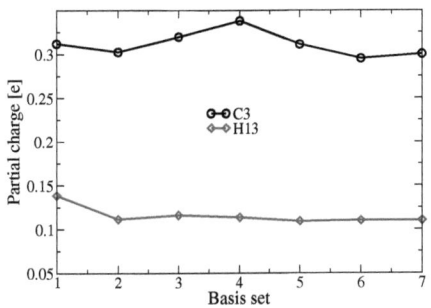

Label	Charge	Label	Charge
N1	-0.62	C2	0.30
C3	0.30	C4	0.00
C5	0.00	C6	-0.27
C7	-0.27	C8	0.09
C9	0.09	C10	-0.16
C11	-0.16	C12	-0.15
C13	-0.15	C14	0.40
C15	0.00	H6	0.11
H7	0.11	H10	0.08
H11	0.08	H12	0.11
H13	0.11		

Tab. 5.3.: Charge convergence with respect to basis set. The numbers one through six refer to 6-311G(d), 6-311G(d,p), 6-311++G(d,p), 6-311++G(2d,2p), 6-311G(2df,2pd) and 6-311++G(3df,3pd) respectively. Number seven is the final choice. The table summarizes the final partial charges for the repeat unit of a carbazole macrocycle in multiples of the elementary charge e. For labeling refer to fig. 5.1a.

5.2. Molecular dynamics simulations

The initial configurations for the MD simulations of the macrocycles are based on x-ray data provided by W. Pisula [103]. According to this data the macrocycles are spaced 0.4 nm apart and rotated by 30° with respect to each other along the column. Since the four OH side chains divide the macrocycle into four identical units, every fourth molecule along the column is in the exact same configuration (see fig. 5.7a). The columns are arranged on a hexagonal lattice with a packing distance of 4.7 nm. To obtain stable MD starting configurations, this distance had to be reduced to 4.2 nm by moving the columns closer together. The distance of closest approach between columns is significantly smaller, since the diameter of the rings is approximately 3.8 nm. As a model system, the bulky, branched alkyl side chains reported in ref. [103] were substituted by linear chains four carbons in length [106]. This reduces the steric hindrance in the initial MD structure and thus simplifies equilibration. It also creates space in the macrocycle to insert a central π-system, such as a graphene molecule [106], to form a coaxial cable. Finally, the synthesis of alkyl side chains with modified length or position is straightforward. The OH side chains were left exactly as they were synthesized. A comparison between the synthesized and the simulated ring structure is shown in fig. 5.7. The initial configuration of 4 by 4 by 48 carbazole macrocycles, i.e. 16 columns totaling 224,256 atoms, is shown in fig. 5.8.

All MD simulations were performed using the *Gromacs* program [46] with the velocity Verlet MD integrator and a time step of 1 fs. The runs use a smooth particle mesh Ewald (PME) algorithm [115] with explicit treatment up to $r_c = 1.2$ nm to calculate electrostatics and a cut-off of also 1.2 nm for van der Waals interactions. Initial velocities were generated from a Boltzmann distribution at 300 K. The NPT equilibration run lasted 2.4 ns with Berendsen pressure coupling and the velocity rescaling including stochastic dynamics [51] thermostat. It is intended to stabilize the system at a given pressure. In the course of the simulation, the columns moved even closer together, while neighboring molecules within the column moved slightly further apart. The box dimensions thus changed from initially $x_i = 15.3$, $y_i = 13.0$ and $z_i = 20.2$ nm to the final size of $x_f = 14.7$, $y_f = 12.6$ and $z_f = 21.6$ nm. The convergence of system volume as well

5. Carbazole macrocycle

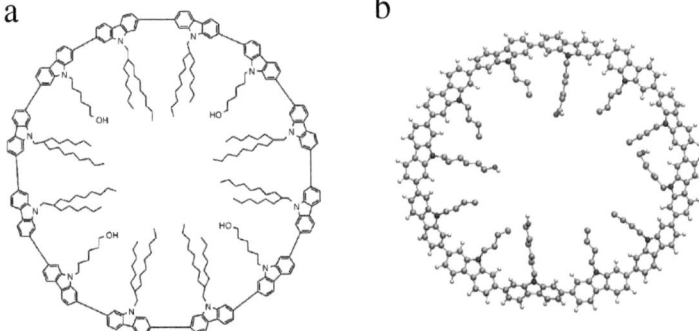

Fig. 5.7.: (a) Structure of synthesized ring taken from [103]. (b) United atom structure with truncated alkyl side chains as used for MD simulations.

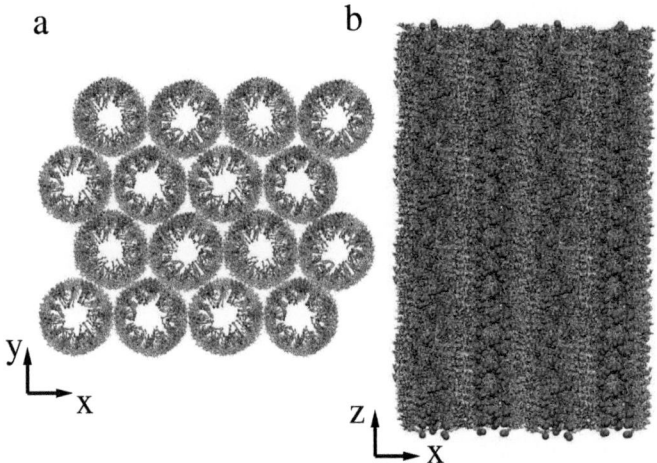

Fig. 5.8.: Initial configuration of carbazole macrocycle columns in (a) top and (b) side view. Oxygens at the end of the side chains are highlighted to illustrate the twisted columnar structure.

as that of the individual box lengths is checked at the end of the NPT run. Potential, kinetic and total energy of the system were conserved, pressure fluctuated around one bar and temperature around 300 K at the end of the run.

Following the NPT simulations, we performed NVT simulations using a box size obtained by averaging over the last few hundred NPT snapshots. Starting positions and velocities were taken from the last NPT frame. The NVT run was first equilibrated for 1 ns using the velocity rescaling thermostat. Then the NVT production run was performed for 20 ps, writing out results every 20 fs. Equilibration was checked by analyzing system temperature, pressure, potential, kinetic and total energy.

A snapshot of the final run is shown in fig. 5.9. In comparison to the initial configuration (see fig. 5.8) the spiral ordering of the oxygen atoms at the end of the side chains is lost, indicating side chain motions as well as rotations of macrocycles with respect to each other. Also, the macrocycle positions deviate from their initially perfect columnar arrangement by fluctuating in the xy-plane, allowing a closer approach and thus higher electron density overlap between molecules of neighboring columns.

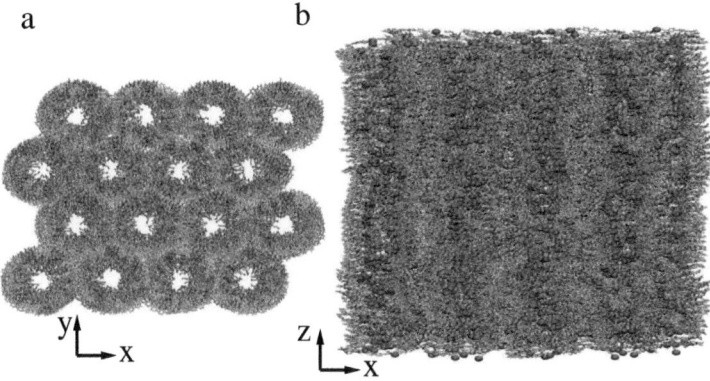

Fig. 5.9.: Final configuration of carbazole macrocycle columns in (a) top and (b) side view at the end of the NVT production run. Oxygens at the end of the side chains are highlighted.

5.3. Charge transport parameters

To simulate the charge dynamics in carbazole macrocycle columns, it is important to determine charge transport units (conjugated segments) in a carbazole macrocycle. Several mechanisms can localize charges: polaron self trapping [116], chemical defects [117], the torsional angle between neighboring monomers [111] and the dynamic disorder in transfer integrals due to thermal fluctuations [74]. The latter mechanism is typically responsible for charge localization in molecular crystals. To treat the conjugation break in a carbazole macrocycle, two limiting approaches are taken. The first assumes that the entire macrocycle is conjugated and thus the charge carrier is spread out over the whole macrocycle. The second assumes that there is no conjugation between the twelve monomer units, i.e. the charge is always localized on one of the monomer segments. In this case, there is also intra-chain transport within the macrocycle. The former approach requires knowledge of reorganization energy and orbitals for a macrocycle, while the latter requires the respective input parameters for a monomer only. In both cases the side chains were assumed not to play a role in the charge transport. The reorganization energy (sec. 4.5.2) was computed based on B3LYP/6-311G(d,p) optimized geometries and wave functions. For the cation we obtained 0.1 eV for the monomer and 0.06 eV for the macrocycle. The transfer integral calculations employ ZINDO orbitals computed based on a B3LYP/6-311G(d,p) optimized geometry. The HOMO and LUMO orbitals of a single carbazole unit as well as the change in transfer integral upon translating two initially cofacial units in x-direction with respect to each other are shown in fig. 5.10. The change in transfer integral upon translation is described by an oscillatory exponential decay. Clearly, there is no more electron density overlap once the molecules have moved completely apart and it decreases from the closest cofacial configuration up to that point. The oscillation arises from the bonding-antibonding pattern of the orbitals, which depends on the sign of the wave function. In face-to-face arrangement the transfer integrals favor hole transport (HOMO-HOMO electronic coupling), but for certain displacements electron transport becomes favorable.

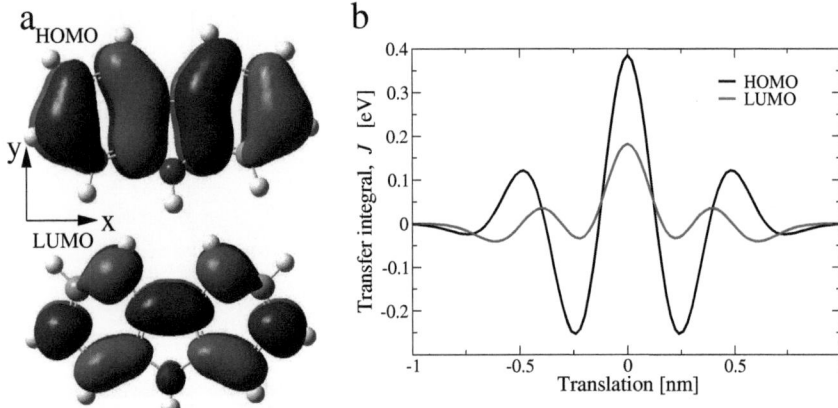

Fig. 5.10.: (a) Shape of HOMO and LUMO for a carbazole monomer. (b) Value of transfer integral for HOMO and LUMO upon translating two cofacially aligned monomers with respect to each other in direction of the arrow indicated in (a). The spacing between the molecules is fixed at 0.36 nm. Modified from [100].

The HOMO orbital of the macrocycle along with the change in transfer integral upon rotation of two planar macrocycles with respect to each other is shown in fig. 5.11. Since the macrocycle consists of twelve monomers, a rotation by thirty degrees reproduces an identical configuration. In the actual MD simulation the rings are never planar, neither are they in their optimized vacuum geometry, where the dihedral angle between monomers is approximately 140°. Nonetheless the calculation illustrates that the transfer integral between macrocycles also favors hole transport and that the bonding-antibonding configurations lead to two minima as well as one maximum, which are as strong as that of the cofacial configuration. There are also positions of zero coupling in between at rotations of 3.75, 11.25, 18.75 and 26.25°. This shows the importance of the azimuthal alignment of the molecules for charge transport.

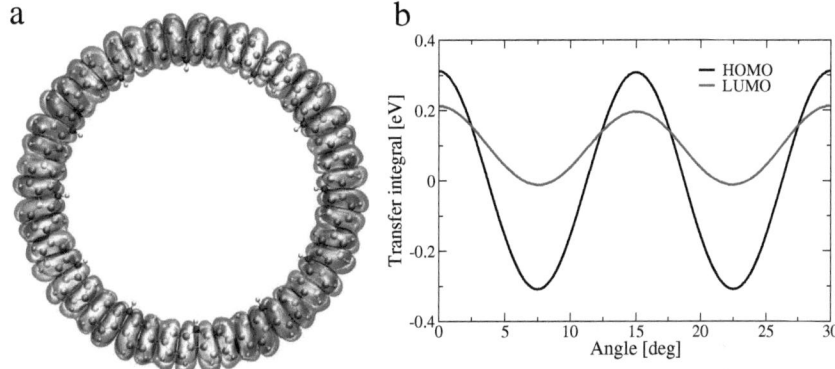

Fig. 5.11.: (a) HOMO of the carbazole macrocycle. (b) Change in transfer integral upon rotating two planar macrocycles without side chains with respect to each other [100].

5.4. Rate-based simulations of charge dynamics

KMC simulations were performed based on the previously determined morphologies using the VOTCA package [95]. The externally applied electric field was chosen to have a magnitude of $E = 10^7$ V/m throughout. It is always aligned parallel to the direction of transport being studied. The dipole moment was computed for configurations optimized in vacuum. It yields approximately 2 Debye parallel to the direction the nitrogen is facing in. In the macrocycle the dipole moment is reduced to 0.03 Debye, because the monomer dipoles facing toward the center of the cycle compensate each other. Note that there is a 140° angle between consecutive monomers in the optimized macrocycle. When neglecting electrostatic disorder the site energy difference reads $\Delta G = e\mathbf{E}\mathbf{r}_{ij}$, where e is the elementary charge, $\mathbf{E}$ the electric field and $\mathbf{r}_{ij}$ the vector connecting the two sites i and j. Since the rings deviate strongly from their equilibrium configuration in the MD runs, results with and without electrostatics are compared. To analyze the effect of different conjugation lengths along the macrocycle the two limiting cases are treated: that the conjugation is spread over the entire macrocycle and that it is broken after each repeat unit.

5.4.1. Orbital mapping and modeling of intra-chain transport

To treat a fully conjugated macrocycle and a cycle composed of twelve conjugated segments, i.e. carbazole monomers, two different orbital mappings are required. In the first case (the full ring mapping), the entire carbazole macrocycle is treated as one hopping site and its orbitals are calculated for an optimized macrocycle with a dihedral angle of 140° between the monomer segments. In the course of the MD simulations this angle deviates significantly from its equilibrium value in contrast to bonds and angles within each monomer, which remain approximately constant. Thus the orbitals of the optimized macrocycle are rotated to match the monomer orientation of each MD snapshot. This is possible since the ZINDO orbitals (sec. A.1.7) are stored as linear combination of atomic orbitals (sec. A.1.3), which allows rotation of the macrocycle orbitals independently from each other based on the orientation of the underlying atoms. In the second case (the monomer mapping), each monomer corresponds to a hopping site and its orbitals are computed for a single carbazole unit, i.e. a monomer saturated with hydrogens, optimized in vacuum. To treat intra-chain transport, i.e. transport along the macrocycle in the monomer mapping, transfer rates ω_{ij} between consecutive monomers i and j along the cycle must be calculated. For this purpose an expression derived from transition state theory [87, 118] is used:

$$\omega_{ij} = \nu \exp\left\{-\frac{(\Delta G_{ij} - \Lambda)^2}{4\Lambda k_B T} - \frac{|J_0 \cos \Psi_{ij}|}{k_B T}\right\}$$

Here ν is a prefactor related to the frequency of the promoting mode and the relevant Franck Condon factor [87]. To obtain reasonable rates, a value of $\nu = 10^{15}$ sec^{-1} is used in accordance with ref. [119]. This ensures that intra-chain rates are fast enough so that charge transport is still limited by inter-chain rates. As in sec. 4.4.2 and eqn. 4.24, ΔG_{ij} is the free energy difference between the charge being located on molecule i and j, Λ is the reorganization energy and $k_B T$ the thermal energy. The intra-molecular transfer integral is approximated by $J_{\text{intra}} = J_0 \cos \Psi_{ij}$, where Ψ_{ij} is the torsional angle between monomers (fig. 5.1d) and J_0 is chosen to have a value of 0.1 eV.

5.4.2. Transfer integral distributions & cut-off radius

The KMC simulations require the specification of a cut-off radius r_c to define nearest neighbor pairs in a simulation. All molecules whose center of mass lies within a sphere of radius r_c around the center of mass of molecule i are treated as nearest neighbors of i. Transfer integrals and corresponding rates are calculated between i and its nearest neighbors ignoring all other molecules in the vicinity. Therefore, if molecules are labelled in order of stacking along the columns and the cut-off radius is large, the transfer integral between molecules i and $i + 2$ is calculated without taking into account the presence of molecule $i + 1$, which lies in between. However, they have little influence on the KMC simulations because the corresponding transfer integrals and hence rates are low. This is shown in fig 5.12a by the distinct maxima for first and second order neighbors in the distribution of the logarithm of the square of the transfer integrals. A cut-off radius of $r_c = 0.7$ nm includes only nearest neighbors, while $r_c = 1.2$ nm shows a peak due to second order neighbors (see fig. 5.12b).

Since the side chains for the carbazole macrocycles face inwards, a close approach between cycles of neighboring columns becomes possible. This enables charge transport perpendicular to the stacking direction of the columns, i.e. in the xy plane. To account for transport in this direction, the cut-off has to be increased significantly beyond the distance between nearest neighbors in a column. Concomitantly, this would mean the inclusion of a large number of neighbors in the z direction which do not contribute to the charge transfer process. To avoid the unnecessary calculation of transfer rates for these higher order neighbors, an additional cut-off r_z along the z-direction is introduced. It is set to $r_z = 0.6$ nm for all simulations and removes neighbor pairs where $r_z < 0.6$ nm despite the fact that they lie within the cut-off radius r_c. Both 1D and 3D transport are studied in the full ring mapping. In case of 1D transport,

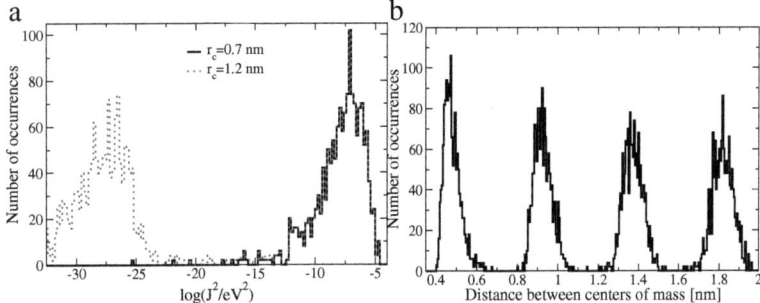

Fig. 5.12.: (a) Distribution of the logarithm of the square of the transfer integral for cut-off radii of $r_c = 0.7$ and 1.2 nm. The former is the cut-off radius for the one-dimensional KMC runs. (b) Distribution of distances between centers of mass for the full ring mapping.

$r_c = 0.7$ nm is chosen based on the distribution of center of mass distances shown in fig. 5.12b, which splits into clearly separable peaks corresponding to first, second, third and higher order neighbors. For 3D transport r_c is set to 3.8 nm. In this case in addition to the two neighbors in the same column, each molecule may have up to twelve additional neighbors in the xy plane. In hexagonal ordering each column is surrounded by six neighboring columns in hexagonal packing and if column j is also shifted along the z-axis with respect to column i, two neighbors become accessible. The resulting percolation networks for one- and three-dimensional transport are illustrated by connectivity graphs in fig. 5.13. Grey spheres represent the centers of mass of each macrocycle, bonds are drawn between molecules for which coupling was calculated. The width of the bond is proportional to the transfer integral, the color represents the sign with blue being positive and red negative.

In case of the monomer mapping there are intrachain neighbors to allow transport within the ring, whose coupling depends solely on the dihedral angle between consecutive segments and not on their distance or the cut-off radius r_c. In addition, interchain transfer integrals corresponding to hops from a monomer on ring i to the closest monomer on ring $i + 1$ are taken into account. As for the full ring mapping $r_z = 0.6$ nm while r_c is set to 0.9 nm. The latter choice is made because, on the one hand, the dependence of mobility on cut-off radius reaches a plateau at 0.9 nm (see fig. 5.14a). On the other hand, it is not possible to distinguish between intra-columnar and inter-columnar neighbors based on the center of mass distribution of the monomers, as shown in fig. 5.14b. Thus, only 3D transport is treated for the monomer mapping. A choice of $r_c = 0.9$ nm permits up to eight neighbors per monomer, i.e. two for each macrocycle above and below along the column and two for each of the two closest macrocycles in the closest column.

5.4.3. Charge mobility

In all cases the charge mobility was calculated using velocity averaging for 10^{-7} seconds based on the last 100 frames of the MD production run with 10 different starting positions per frame. To show the influence of static disorder, additional KMC simulations were performed for an ideal configuration based on x-ray data. To this end, the rings as optimized in vacuum were

5. Carbazole macrocycle

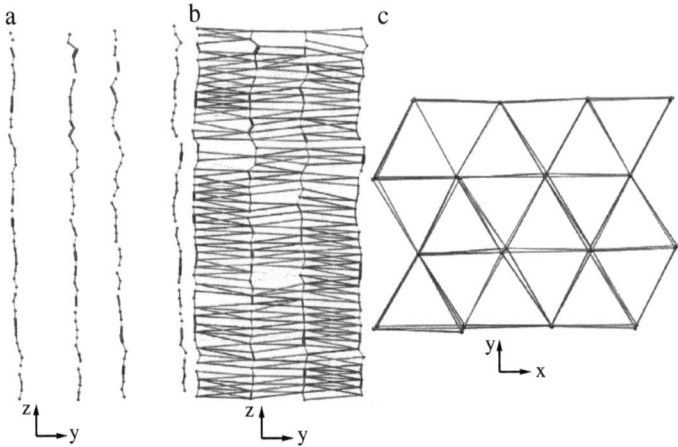

Fig. 5.13.: Connectivity graphs showing only nearest neighbor connections for the full ring mapping in case of (a) $r_c = 0.7$ nm and interconnections between columns for (b) and (c) where $r_c = 3.8$ nm. The distance in z direction between neighbors never exceeds $r_z = 0.6$ nm. For clarity only the first four columns are displayed in (a) and (b) and the upper two layers in (c).

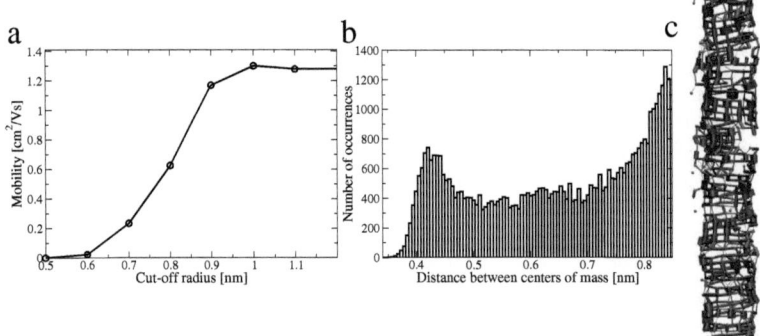

Fig. 5.14.: (a) Dependence of mobility on cut-off radius for the monomer mapping. (b) Distribution of distances between centers of mass. (c) Connectivity graph for one column in monomer mapping. Bonds are not displayed if coupling between neighbors is below 10^{-6} eV.

stacked into hexagonally ordered columns. Within the columns consecutive molecules are rotated by 30° with respect to each other. In this case KMC runs were averaged for 100 different starting positions. Finally, we have also calculated the mobility for the equilibrated morphologies taking into account electrostatic disorder. The resulting hole mobilities for the ideal and the equilibrated system with and without electrostatic disorder are summarized in tab. 5.4.

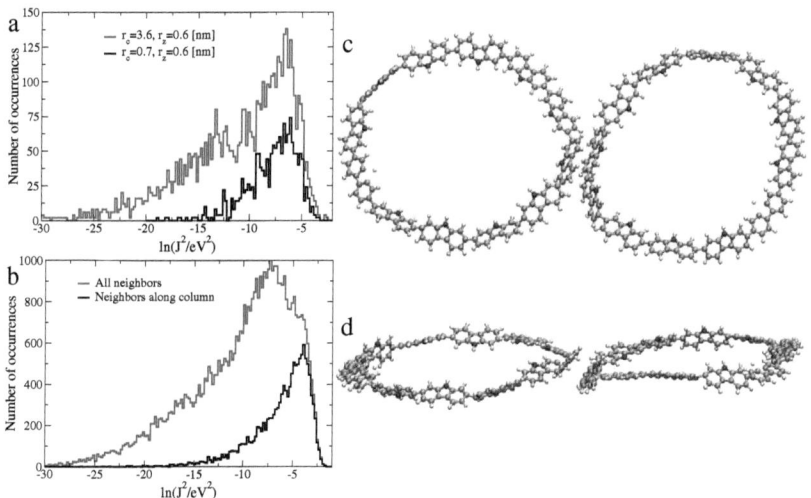

Fig. 5.15.: Distribution of the logarithm of the square of the transfer integral (a) for the fully conjugated and (b) for the monomer mapping. The magnitude of intra-columnar transfer integrals ($r_c = 0.7$ nm) is compared to that of inter-columnar neighbors (additional neighbors for $r_c = 3.8$ nm). In (c) and (d) the alignment of a neighboring pair with a high coupling of $J \geq 0.01$ eV is displayed in top and side view, respectively.

In the one-dimensional case using the fully conjugated mapping, the mobility is more than an order of magnitude lower ($1.2 \cdot 10^{-3}$ cm^2/Vs) than in the equilibrated case (0.05 cm^2/Vs). In the ideal system neighboring pairs with low coupling exist because the quantum chemically equilibrated cycle is not planar. Thus while planar cycles have their maximum transfer integral at a rotation of 30° with respect to each other, in case of the optimized cycles the transfer integral also depends on the respective orientation of monomers of consecutive cycles and may thus be quite low even at a 30° angle. In the equilibrated morphologies, the monomer orientations of consecutive cycles align better. However, the mobility varies by two orders of magnitude (between 10^{-1} and 10^{-3} cm^2/Vs) for different snapshots. The large variation is expected, because a single low transfer integral along a column reduces the mobility in a one-dimensional system significantly, since every charge has to travel across the pair with low coupling and therefore the mobility depends on the lowest transfer integrals in the system. This was already shown for hexabenzocoronene in refs. [113, 92, 94].

In case three-dimensional transport is treated with the fully conjugated mapping, charge mobility increases from 0.05 to 0.5 cm^2/Vs for the equilibrated system and differences in mobilities between frames become negligible. Transport along the z-direction, i.e. in stacking direction of

5. Carbazole macrocycle

	Direction	full ring (1D)	full ring (3D)	monomer
equilibrated	x	-	1.195	0.566
	y	-	1.176	0.554
	z	0.052	0.465	1.165
ideal lattice	x	-	0.187	0.573
	y	-	0.200	0.550
	z	$1.15 \cdot 10^{-3}$	0.366	1.839
with electrostatic disorder	x	-	0.648	0.074
	y	-	0.686	0.121
	z	0.039	0.261	0.240

Tab. 5.4.: Hole mobilities (in $cm^2V^{-1}s^{-1}$) obtained by velocity averaging. Results for the fully conjugated cycle are compared to a cycle composed of twelve conjugated monomer segments. For the full ring two different cut-offs are used: $r_c = 0.7$ nm corresponding to coupling only between nearest neighbors along the column (1D) and $r_c = 3.8$ nm allowing for inter-columnar and thus three-dimensional transport (3D). The table also compares results for the equilibrated MD configurations with and without taking into account electrostatic disorder with results for the ideal crystal lattice based on x-ray data.

the columns, is also of similar magnitude in the equilibrated and the ideal case. The improvement is due to the fact that in three dimensions any defect in a column can be circumvented. Thus, the low transfer integrals in a column no longer decrease the mobility. Transport in the xy plane is weaker in the ideal than in the equilibrated morphology, but in both cases it is of the same order of magnitude as transport in the z direction. For the monomer mapping, the mobility in the ideal case is higher than in the fully conjugated mapping in all directions. Upon equilibration, transport along the z-axis slightly decreases, but remains higher, while mobility in the xy plane remains the same and thus weakens compared to the fully conjugated mapping.

The difference between inter and intra-columnar coupling is illustrated by the distributions of the logarithm of the square of the transfer integral shown in fig. 5.15a and 5.15b. The bulk of high transfer integral values is found along the column, but while inter-columnar couplings form the tail of the distribution, a few of them reach values comparable to the highest intra-columnar transfer integrals. The latter is more often the case for the fully conjugated cycles (a) than for the monomers (b). In both cases high inter-columnar transfer integral values are found when the rotation of the closest monomer units of neighboring macrocycles is such that it leads to almost cofacial coupling between them. An example of such a neighboring pair with a coupling of $J = 0.011$ eV is shown in fig. 5.15 in side (c) and top (d) view.

The different results obtained for monomer and fully conjugated mapping are rationalized as follows. The higher mobility in the xy-plane for the full ring mapping is due to the fact that when a charge carrier hops in the xy-plane it travels more than 3.5 nm in the fully conjugated case while it travels less than 0.9 nm in the monomer case. In the z-direction the monomer mapping yields higher mobilities because the fast intrachain transport allows the charge carrier to choose the monomer with the highest coupling to hop to the next macrocycle. In contrast, the coupling between two fully conjugated cycles always takes into account the unfavorably aligned monomers as well. As shown in fig. 5.15, more high transfer integral values are present in the monomer (b) than in the fully conjugated mapping (a). Overall, the different mapping schemes do not significantly influence the mobility.

Influence of electrostatic energetic disorder

Energetic disorder is known to decrease the mobility by several orders of magnitude in compounds with a strong dipole moment. Because of its symmetry the carbazole macrocycle has a low dipole moment of 0.03 Debye. Its constituent monomers, however, show a dipole moment of 2 Debye. Thus we decided to take into account energetic disorder due to electrostatics. To compute the energetic disorder, we used the partial charges calculated using CHELPG based on the geometry of a carbazole macrocycle in vacuum optimized by B3LYP/6-311G(d,p) in both charged and neutral state. For the monomer mapping we took the partial charges of a single unit in vacuum and included the charge of the two hydrogens added for saturation (H8 and H9) in that of the carbons forming the bond in the macrocycle (C8 and C9 in fig. 5.1). The electrostatic site energy difference contributing to ΔG_{ij} in the Marcus rate was thus computed as described in sec. 4.5.4.

Introduction of electrostatic disorder leads to a reduction of mobility by a factor of two in case of the full ring mapping (0.5 to 0.25 cm^2/Vs in z-direction), while it reduces the mobility by more than a factor of five in case of the monomer mapping (1.8 to 0.25 cm^2/Vs in z-direction). Overall, the mobility along the columns becomes identical in both mappings, while the mobility in the xy plane is now significantly better for the fully conjugated (0.6) than for the monomer mapping (0.1 cm^2/vs). Due to the higher dipole moment of a single carbazole unit, it is not surprising that the monomer mapping is more affected by introduction of electrostatic disorder than the full ring mapping. However, in both cases the reduction in mobility is less than an order of magnitude.

5.4.4. Conclusions

In this chapter we derived all force field parameters necessary to describe carbazole with alkyl side chains attached as well as polycarbazoles, such as the macrocycle. With the guiding help of x-ray data this allowed us to study the local ordering of carbazole macrocycles. Charge transport simulations were performed using two different charge transport units, namely the entire macrocycle and the carbazole monomer. Comparison of the corresponding transfer integral distributions and mobilities revealed only minor differences indicating that in our model intra-chain transport is so fast that it is almost equivalent to full conjugation.

When electrostatic disorder is taken into account, the mobility is reduced by a factor of two to five depending on the model. Due to the strong dipole moment of a single carbazole monomer, a decrease in mobility by orders of magnitude could have been expected. However, the symmetry of the macrocycle remains throughout the MD simulations and thus makes the influence of electrostatic disorder almost negligible.

The most interesting property of this columnar system is that it conducts in three-dimensions, achieved by synthesis of macrocycles with side chains facing inwards. The introduction of intercolumnar transport leads to an increase in mobility by at least an order of magnitude compared to solely intra-columnar transport, because it provides pathways to circumvent neighbors with low coupling along a column. So far there are no experimental measurements to support this but, provided the supramolecular arrangement is achieved, local mobilities as measured by PR-TRMC (see sec. 2.2.2) are expected to be on the order of 0.1 cm^2/Vs for such systems. Macrocycles with side chains on the outside will prevent three-dimensional transport and are thus expected to have mobilities on the order of 10^{-3} cm^2/Vs.

6. [1]Benzothieno[3,2-b]benzothiophene (BTBT)

One of the main reasons why organic materials show promise to replace their inorganic counterparts is that devices can be built using solution processing [120, 121]. The drawback is that solubility requires the addition of flexible side chains, which are bound to negatively influence the order in the resulting crystal morphology. In this chapter we intend to study the influence of side chains on a self-assembled crystal morphology created by solution processing. Flexible side chains introduce static - in comparison to the speed of charge carriers - and dynamic disorder, which requires a sophisticated percolation path analysis to predict charge transport properties.

Presently, multiple methods are available to calculate charge transport in such systems depending on the transport regime [63, 64, 66, 82, 93, 94]. It has been shown in refs. [73, 74] that for organic crystals the thermal fluctuations of transfer integrals between neighboring molecules is of the same order of magnitude as their average values. This allows us to assume that the electronic wave function is not completely delocalized over the entire structure at ambient conditions, which excludes the notion of band-like transport. If it is still spread out over a small number of molecules, transport ought to be treated as diffusion limited by thermal disorder. This can be achieved using semi-classical dynamics based on a model Hamiltonian with interacting electronic and nuclear degrees of freedom [75]. If, however, nuclear dynamics are much slower than the dynamics of the charge carriers and coupling is weak, transport should be described by a Hamiltonian with static disorder based on the electronic density of states and on hopping rates between localized states. Charge transport may then be computed using rate equations, for example by kinetic Monte Carlo based on Marcus rates [93, 94]. Up to now, it is still not clear which method is most suitable for the study of partially disordered semiconductors, since it is unknown how much disorder is truly present in these systems. Therefore, a comparison of charge transport characteristics based on semi-classical dynamics as well as on rate equations for a relevant test case may elucidate which transport regime is more appropriate for crystals with significant disorder due to side chain motions.

As a test system, we chose [1]Benzothieno[3,2-b]benzothiophene (BTBT, see fig. 6.1), which was initially used as a high-performance molecular semiconductor for vacuum-processed OFETs, where it showed excellent stability and lifetime under operation in air [122, 123]. By introduction of alkyl side chains in the long-axis direction of the core, which ought to facilitate lateral intermolecular interaction, BTBT could be made sufficiently soluble for solution-processing. A comparative study of OFETs created by spincoating BTBT with alkyl side chains of different length on an Si/SiO_2 substrate showed self-organization into crystalline structures as well as excellent device mobilities [123]. The highest mobility ranging between 1.20 and 2.75 cm^2/Vs was found for C13 side chains. For the simulations we use C12 side chains, since the corresponding x-ray structure is readily available in the supporting information of ref. [123] and with values ranging between 0.5 to 1.7 cm^2/Vs the mobility is not significantly below that of the C13 side chains. Thus the crystal structures are likely to be similar as well. The chemical structure and unit cell of BTBT with C12 side chains are shown in fig. 6.1a and b. The unit cell has P1 symmetry with the main axes $a = 5.864$ nm, $b = 7.740$ nm and $c = 37.910$ nm. The corresponding angles are $\alpha = 90.00°$, $\beta = 90.59°$ and $\gamma = 90.00°$.

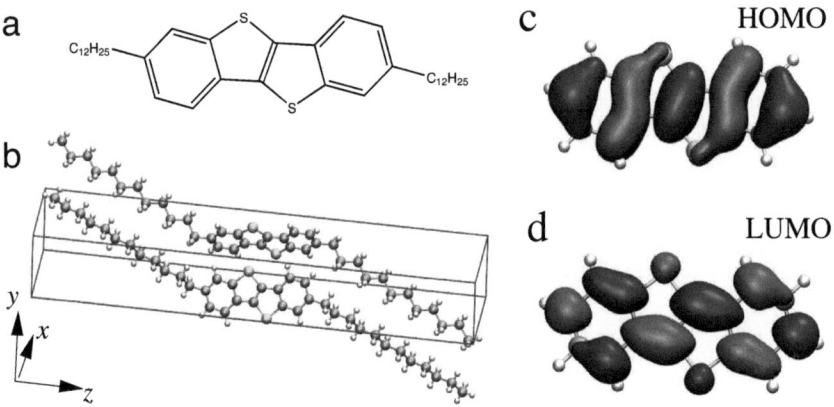

Fig. 6.1.: (a) Chemical structure of [1]Benzothieno[3,2-b]benzothiophene (BTBT) with C12 side chains. (b) Unit cell of the BTBT-C12 crystal. Taken from [124]. The HOMO and LUMO calculated using ZINDO are illustrated in (c) and (d).

We first present the MD simulations performed for BTBT-C12, to be followed by charge transport simulations using rate equations (sec. 6.2) and semi-classical dynamics (sec. 6.3). The side chain motions are shown to give rise to static disorder in the transfer integral distributions, which leads to a modification of the percolation pathways in the system. Finally, a comparison of the two different simulation approaches is given.

6.1. Force field and MD simulations

As before, the first step of an MD based study of charge transport is the determination of suitable force field parameters for the MD run. Since no complete set of parameters is available for BTBT in the OPLS force field, we first assign the atom types and force constants by using analogies to thiazole, furan, benzene and indole, which are present in OPLS. The rest of the parameters are obtained by fitting to quantum chemical potential scans, as described in sec. 3.3.2. Detail regarding the BTBT force field are given in the appendix A.3.1). MD simulations were performed using the *Gromacs* package [125]. The initial configuration of 12 by 12 by 6 molecules in x-, y- and z-direction was built based on the above mentioned experimental unit cell. The smaller number of molecules along the z-direction was chosen since this is the direction of the side chains where no charge transport is assumed to occur. The system was equilibrated for 250 ps in NPT ensemble, then one ns in NVT ensemble and finally production runs were performed in NVT ensemble with snapshots taken every 20 fs for a total of 20 ps. The short production run time is due to the high frequency of the output required to compute the time autocorrelation function of transfer integrals for semi-classical dynamics, which will be discussed in sec. 6.3. The rate equation approach requires only a single MD snapshot in principle, although multiple snapshots are useful for statistical reasons. An additional production run of one ns in NVT was performed to confirm the slow side chain motions. All simulations were performed at a temperature of 300 K using Particle Mesh Ewald with a grid spacing of 0.12 nm and explicit treatment up to 1.1 nm for the Coulomb interactions and a cut-off of 1.1 nm for the van der Waals interactions. Initial pressure equilibration was done with the Berendsen barostat. The Nose-Hoover thermostat was used for temperature equilibration and velocity rescaling [46] for the production runs.

6.2. Charge transport using Marcus rates

As for carbazole (sec. 5.3) we treat the side chains as insulators and calculate the relevant HOMO and LUMO orbitals solely for the BTBT core, i.e. with the side chains being replaced by hydrogen atoms. The core geometry was optimized using B3LYP/6-311G(d,p) and the orbitals were calculated using ZINDO. Both HOMO and LUMO are non-degenerate and visualized in fig. 6.1c and d. The internal reorganization energy Λ for cations and anions is 0.229 eV and 0.303 eV, respectively. Since all molecules in the system are identical, there is no diagonal disorder in the HOMO energies. Electrostatic and polarization contributions to the energetic disorder are ignored, due to the small dipole moment of a single optimized molecule of 0.005 Debye. Thus $\Delta G_{ij} = e\mathbf{E} \cdot \mathbf{r}_{ij}$, where $\mathbf{E}$ is the applied electric field, $\mathbf{r}_{ij}$ is the vector connecting molecules i and j and e is the elementary charge. Hopping rates were calculated using Marcus theory as described in sec. 4.4.2.

Charge mobility was calculated using the time of flight, velocity averaging and Einstein relation approaches (see sec. 4.4.4). Velocity averaging is the most accurate, because it is straightforward to obtain the mobility from the resulting velocity while the other two approaches require a fitting procedure to extract the mobility. For velocity averaging and time of flight an electric field with magnitude of $E = 10^7$ V/m was applied. The field direction was always aligned parallel to the transport direction. The results obtained for mobilities in x- and y-direction as well as diagonally across the xy plane are shown in table 6.1. Agreement between the methods is very good, indicating that the chosen field strength is in the low field regime. Due to our treatment of the side chains as insulators, there is no transport along the z-direction, but the x- and y-direction are similarly suited for transport. This is also supported by the xy projections of two typical diffusion paths lasting 10^{-2} seconds shown in fig. 6.2a.

Direction	TOF	$\langle v \rangle$	Einstein relation	crystal
x	0.089	0.0795	0.0826	0.377
y	0.099	0.0813	0.0734	0.224
xy	-	0.0884	0.0692	0.302

Tab. 6.1.: Hole mobilities (in cm^{2}V^{-1}s^{-1}) obtained by the different methods: time of flight (TOF), velocity averaging, $\langle v \rangle$, and diffusion (Einstein relation). The last column is the mobility of the perfectly ordered crystal obtained using velocity averaging.

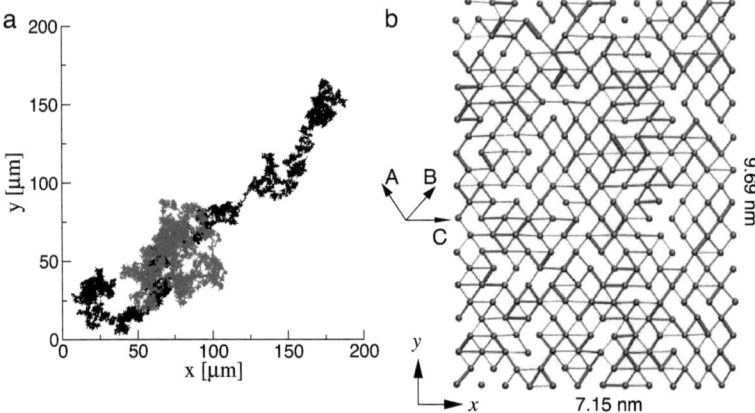

Fig. 6.2.: (a) Diffusion pathways of charges after 10^{-2} sec for two different MD snapshots. (b) Connectivity graph in one of the xy-planes. The spheres indicate the centers of mass of the molecules, bonds are displayed if the connecting transfer integrals exceed 0.005 eV. The thickness of the bonds is proportional to the magnitude of the transfer integrals, the color represents the sign with blue being positive and red negative. Note the difference between the size of the slab and the much larger diffusion range of the charge shown in (a).

Mobilities of crystalline organic semiconductors are often estimated based on a static unit cell [126, 24]. Thus, we also compare the ensemble-averaged mobility to the one calculated for an ideal, perfectly ordered crystal. In the ideal case, velocity averaging predicts an about four times higher mobility ($\sim 0.3\,\mathrm{cm^2V^{-1}s^{-1}}$) with a noticeable preference for the x- over the y-direction (see again tab. 6.1). To understand the discrepancy between the equilibrated and the perfectly ordered case we analyzed the transfer integral distributions. The total distribution can be split into subdistributions corresponding to the three directions responsible for charge transport, A, B and C, i.e. the three nearest neighbor pairs shown in the inset of Fig. 6.3. For the ideal crystal they are all approximately equivalent: $J_A = J_B = J_C = 0.023$ eV. Since a hop in the C direction corresponds to a step of 0.58 nm along the x-axis and no motion in y-direction, but a hop in A or B direction means only 0.39 nm along the y-axis and still 0.29 nm along the x-axis, it is clear that transport along the x-axis is preferred in the ideal crystal. The equilibrated system (see fig. 6.3b) on the other hand shows a narrow distribution centered around 0.007 eV for the C direction and two almost identical distributions for directions A and B, which are extremely broad and centered around zero. At first glance one might expect only weak transport in the C (or x) direction. This perception is misleading, however. An analysis of the transfer integral variation with time $J(t)$ for multiple different molecule pairs shows that the distributions are narrow. But depending on the pair under investigation they are centered around values ranging from -0.06 to 0.06 eV as depicted in fig. 6.4c.

This indicates that the wide distributions are due to static disorder and occur when we average over an ensemble of molecules. The static disorder in the system can be explained by the irregular displacement of the molecules along the z-axis, i.e. in the direction of the side chains as depicted in fig. 6.4b. Indeed, as is shown in the inset of fig. 6.4c even small displacements

6.2 Charge transport using Marcus rates

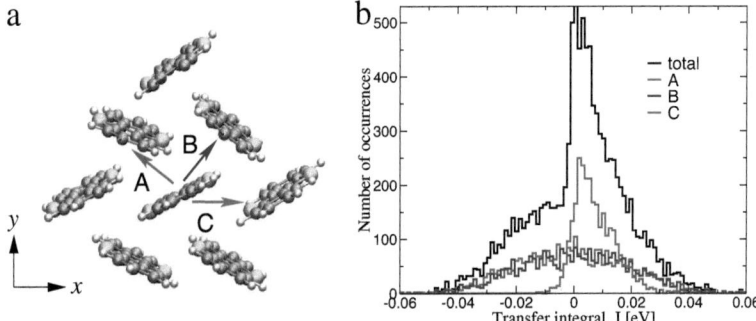

Fig. 6.3.: (a) The neighboring pairs corresponding to the three main transport directions A, B and C with average connecting vectors of $(-0.29, 0.39, 0.0)$, $(0.29, 0.39, 0.0)$ and $(0.58, 0.0, 0.0)$ in nm. (b) Distribution of transfer integrals in a single snapshot composed into these directions.

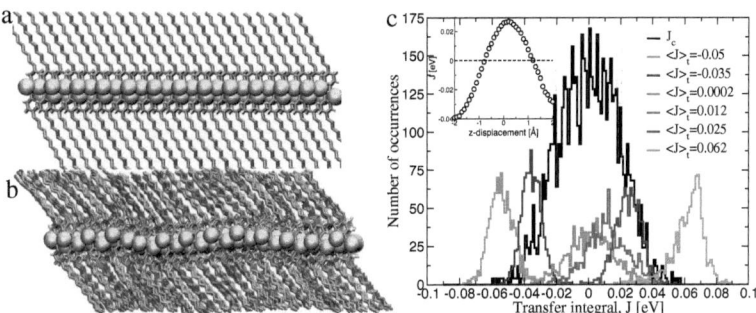

Fig. 6.4.: (a) Ideal crystal alignment. (b) Crystal after equilibration illustrating the displacement in z-direction due to slow fluctuations of the soft side chains. (c) Distribution of transfer integrals in the A and B direction with respect to the configuration (J_c) of a single snapshot, i.e. 11378 pairs, as well as time distributions of the transfer integral for different pairs with different average transfer integrals $<J>_t$, i.e. 1000 time steps of 20 fs each. The inset shows the dependence of J on the displacement of neighbors along the z-axis.

79

6. [1]Benzothieno[3,2-b]benzothiophene (BTBT)

in the z-direction can lead to very large changes in the transfer integral. Due to soft side chains the dynamics of these displacements is very slow (> 100 ps), which leads to practically static disorder of the transfer integrals and correspondingly the rates in the system. The speed of the side chain dynamics was confirmed by comparing the $J(t)$ distributions for the 20 ps and the 1 ns run, showing no significant broadening of the distributions despite the much longer run time.

The importance of static disorder can be visualized with the help of a connectivity graph, which is shown in fig. 6.2b. Here the spheres indicate the centers of mass of molecules and the thickness of the bonds corresponds to the magnitude of the transfer integrals between them. No bonds are drawn if $|J| < 0.005$ eV. The color represents the sign with blue being positive and red negative. Despite the fact that along all three directions neighbors with a coupling of less than 0.005 eV exist, percolation pathways may easily be found. Hence, even though the transfer integral averaged over all pairs is zero for A and B directions, there is always a percolation pathway due to the broadness of the transfer integral distribution of the system. This is solely due to the two-dimensionality of the system: in a one-dimensional case, such as columnar discotics (see sec. 5 and 8), transport would not be possible since it is limited by the tail of small transfer integrals. To further verify the static nature of the disorder, we generated a movie of connectivity graphs versus time for the 1 ns run. It showed some fluctuations, but in general the red and blue patches in the connectivity graph (fig. 6.2b) remained stationary throughout.

An alternative way of visualizing the effect of connectivity on charge percolation is an analysis of the cluster size with respect to transfer integral. Here, a cluster is defined as a set of molecules connected by pathways where all transfer integrals exceed a given transfer integral threshold. The cluster size is defined as the number of molecules in the largest cluster in percent of the system size. For BTBT, in case the transfer integral threshold is zero, this corresponds to 16.67 % or 1/6 of the total system size, since there are six independent planes in the simulation box. As depicted in fig. 6.5 the cluster size remains stable until it rapidly breaks starting at approximately 0.015 eV. This value would be an appropriate choice as effective coupling to describe the system's charge dynamics.

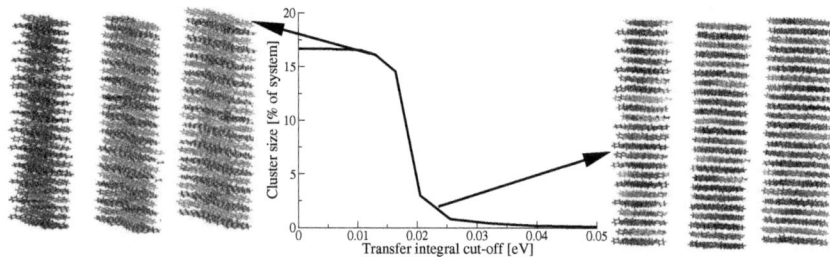

Fig. 6.5.: Dependence of cluster size on transfer integral threshold. The largest clusters within three BTBT planes are visualized for a threshold of $J = 0.015$ and 0.025 eV.

6.3. Charge transport using semi-classical dynamics

Semiclassical dynamics can be used to study the effect of the coupling between charge dynamics and nuclear motion. The basic idea is to build a model Hamiltonian by analyzing the molecular dynamics of the system performed using force-field simulations (see sec. 4.3.1). The model Hamiltonian contains a one-dimensional array of molecules, with nearest neighbor coupling modulated by classical nuclear displacements.

There are a few additional input parameters required for the definition of this Hamiltonian (eqn. 4.11). For the high-frequency mode a single effective mode with $\omega^{(1)} = 1400\,\text{cm}^{-1}$ and mass $m^{(1)} = 6\,\text{amu}$ is chosen [75]. The Holstein coupling parameter $\lambda^{(1)}$ is linked to the reorganization energy Λ of the molecule via $\lambda^{(1)} = \omega^{(1)} \sqrt{m^{(1)} \Lambda}$. Using the reorganization energy for cations from above leads to $\lambda^{(1)} = 37727\,\text{cm}^{-1}$. The average transfer integral is chosen to be the same as for the ideal x-ray structure ($J = 0.023\,\text{eV}$) and the standard deviation ($\sigma_J = 0.010\,\text{eV}$) is taken from the time evolution of the transfer integral for neighboring pairs of type A and B with an average $J(t)$ comparable to that in the ideal crystalline case. The fluctuation of the transfer integral with time along with its autocorrelation for pairs of type A is displayed in figure 6.6a. The fluctuations were analyzed for the trajectory with a stepsize of 20 fs and a total of 1000 steps. The typical vibrational frequency of the molecules is found by calculating the Fourier transform of the autocorrelation of the time-dependent transfer integral $< J(0)J(t) >$, which is shown in figure 6.6b for the three directions of interest, averaged over five neighboring pairs each. The resulting slow vibrational frequency is roughly $\omega^{(2)} = 25\,\text{cm}^{-1}$. This allows calculation of the Peierls coupling constant via $\alpha^{(2)} = \sigma_J m \omega^{(2)} / \sqrt{2 k_B T}$, which gives $\alpha^{(2)} = 627.1\,\text{cm}^{-1}/\text{Å}$. All parameters of the SCD Hamiltonian are summarized in table 6.2.

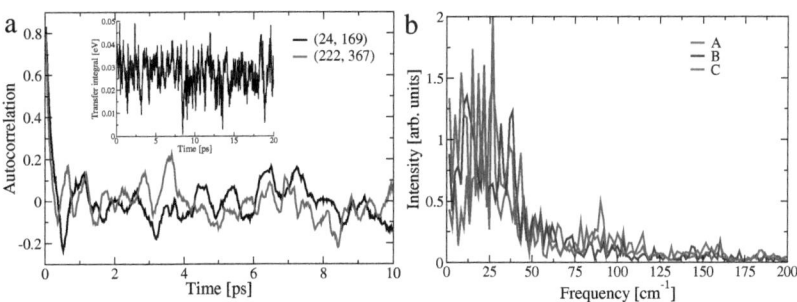

Fig. 6.6.: (a) Time autocorrelation function of the transfer integral for two pairs of type A with $J(t)$ displayed in the inset. (b) Fourier transform of the autocorrelation function for all three directions averaged over five different pairs.

The time evolution of the wave function and the nuclear coordinates is evaluated numerically [127] for 7.5 ps with a time step of 0.0125 fs for a system of 600 molecules under periodic boundary conditions (sec. 4.3.1). The initial wave function is taken from the electronic Hamiltonian eigenfunctions at $t = 0$, and the initial positions and velocities are taken from a Boltzmann distribution at the simulation temperature of 300 K.

The time evolution of the electronic probability distribution along with the resulting mean square displacement and the fit to obtain the diffusion coefficient are shown in figure 6.7. The

6. [1]Benzothieno[3,2-b]benzothiophene (BTBT)

	value	description
J	$285.0\,\text{cm}^{-1}$	average transfer integral
$\lambda^{(1)}$	$37727\,\text{cm}^{-1}/\text{Å}$	Holstein electron-phonon coupling
$\omega^{(1)}$	$1400\,\text{cm}^{-1}$	frequency of mode (1)
$m^{(1)}$	6 amu	mass of mode (1)
$\alpha^{(2)}$	$627.1\,\text{cm}^{-1}/\text{Å}$	Peierls electron-phonon coupling
$\omega^{(2)}$	$25\,\text{cm}^{-1}$	frequency of mode (2)
$m^{(2)}$	576 amu	mass of mode (2)

Tab. 6.2.: Parameters used in the SCD Hamiltonian for BTBT with $C_{12}H_{25}$ side chains.

mobility computed from the Einstein relation $\mu = eD/k_B T$ is $\mu = 1.48\,\text{cm}^2\text{V}^{-1}\text{s}^{-1}$. This value is more than five times larger than the mobility of an ideal crystal with rates given by Marcus' theory, and is due to charge delocalization.

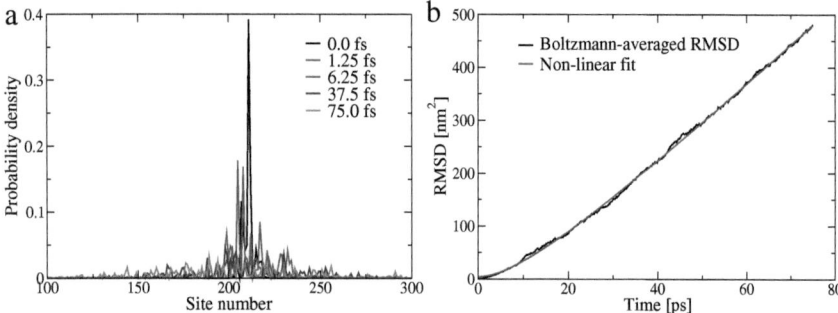

Fig. 6.7.: (a) Spread of the probability distribution. (b) Corresponding Boltzmann-averaged mean squared displacement (MSD) with non-linear fit used to obtain the diffusion coefficient.

The clear advantages of a semiclassical dynamics approach are that it is valid for intermediate coupling strengths ($J \propto \Lambda$) and it includes non-Condon effects (time dependency of the intermolecular coupling). From a practical point of view, however, it can only be applied for short timescales (< 10 ps). Hence, SCD cannot be used for the simulation of mesoscopic disordered or partially ordered morphologies, where slow modes with up to 10 ms relaxation times are easily achievable. In addition, the constructed model Hamiltonian is one-dimensional, i.e. the connectivity of the lattice is ignored, and all molecules are considered to be equivalent, in other words the modulation of the intermolecular coupling and the site energy is the same. Unfortunately, a model that combines the effect of strong coupling [128], time fluctuations [129] and dynamic disorder, that is, a model that strikes a balance between short and long timescale descriptions of the system's quantum dynamics and additionally is capable of treating transport in three dimensions, is currently unavailable and is the main challenge in the field.

In spite of the limitations, SCD results can be used to identify the range of validity of a hopping model for charge transport. Indeed, in the case of BTBT the average intermolecular coupling is rather strong compared to the reorganization energy and one cannot directly assume complete localization on one site. However, the SCD model predicts that the initial wavefunction is localized over approximately six sites because of a combination of dynamic disorder and polaronic self-trapping.

The SCD Hamiltonian, however, neglects static disorder in the system, because its parameters are mapped on a single pair of molecules. As shown in sec. 6.2 the static positional disorder results in roughly 0.02 eV disorder of transfer integrals, which further localizes charge carriers and, hence, strengthens the hopping picture for charge transport in a self-assembled monolayer of BTBT.

6.4. Conclusions

The combination of molecular dynamics simulations, quantum chemical calculations and three different types of theoretical analysis of charge transport provide several insights on the electric properties of partially disordered molecular crystals.

First, it is clear that the ideal situation with a fixed X-ray crystal structure, as it is often used in literature [126, 24], is not adequate for the description of charge transport. Molecular dynamics simulations of self-assembled conjugated systems furnished with flexible side chains show that structural disorder is an intrinsic part of the self-assembled morphology. This leads to the distribution of charge hopping rates between molecules. The topology of the charge percolating network is then defined by this distribution and significantly influences charge dynamics in the system. The idealized picture based purely on a static unit cell does not take this into account and hence gives a misleading impression of the charge dynamics, overestimating the mobility by an order of magnitude. A completely different impression is obtained by considering only effective coupling parameters such as the average transfer integral, which leads to a significant underestimation of the transport properties of the crystal.

Second, one can see a clear advantage of self-assembled layers conducting in 2D rather than self-assembled discotic liquid crystals conducting in 1D. In 2D the presence of the disorder does not disable charge transport, it only modifies the connectivity of the network of molecular sites. In the 1D case off-diagonal disorder introduces low rates in the distribution of charge hopping rates. Since the tail of slow rates defines the mobility along conductive wires, disorder always decreases the charge carrier mobility in 1D systems. This is a very strong argument in preference of 2D conductors (self-assembled monolayers) as compared to discotic liquid crystals.

By comparing two model situations with and without dynamical disorder we also conclude that, for such systems, dynamical disorder has a much weaker effect on charge mobility than static disorder.

Finally, the fact that both rate-based and semiclassical simulations tend to overestimate the mobility of compounds, we are inclined to doubt that the $1.1\,cm^2/Vs$ measured by Ebata et al. [123] is an easily reproducible value. This is also supported by FET mobility measurements of BTBT performed by Yeon Sook Chung and Prof. Do Yeung Yoon, who obtained $0.22 \pm 0.05\,cm^2V^{-1}s^{-1}$ [124].

7. Organic Crystals

In the field of organic electronics several classes of materials are studied, from π-conjugated polymers with aromatic backbones to molecular materials. Both are soluble and can be printed on large scale. Polymers naturally have the better mechanical properties, but structural defects often result in carrier traps and limit the device performance. Organic molecular crystals are advantageous due to their well-ordered structure and ease of purification [130, 131]. Most molecular crystal layers are created by vapor deposition techniques such as molecular beam epitaxy [132], which allows the creation of almost defect-free layers. The resulting well-ordered structures yield high mobilities, but processing is too costly for industrial application. However, it has been shown that soluble derivatives of highly conductive molecular crystals such as pentacene also assemble into well-ordered layers with mobilities above $1\,\mathrm{cm}^2/\mathrm{Vs}$ [133] and thus on the order of amorphous silicon.

To gain a better understanding of what chemical structures and morphologies lead to high mobilities in molecular crystals, we compare rubrene, a molecular single crystal with OFET mobilities of up to $15\,\mathrm{cm}^2/\mathrm{Vs}$ [134, 135, 136, 137] one of the highest in the field, to three other recently synthesized compounds. The first is indolo[2,3-b]carbazole with CH_3 side chains [138], the remaining two are benzo[1,2-b:4,5-b']bis[b]benzothiophene (BBBT) derivatives, one without and one with C4 side chains [139]. In contrast to highly purified rubrene single crystals, which are created by vapor deposition, their structures are obtained by self-assembly after spin coating. By this comparison we intend to understand what makes the structure and morphology of rubrene better suited for charge transport than that of other compounds. In addition the choice of compounds allows us to study the influence of the dimensionality of the charge percolation network on the mobility, because indolocarbazole and BBBT without side chains arrange into 3D transporting morphologies, while rubrene is a 2D and BBBT with C4 side chains a 1D transporting system.

As in the previous chapter, we also compare charge transport simulations based on the diffusion limited by thermal disorder model with simulations based on a charge localized description combined with rate equations to find the appropriate transport regime for well-ordered molecular crystals, where static disorder should be negligible compared to dynamic disorder. We are particularly interested in how the rate-based approach performs in case of rubrene, where neither the assumption of charge localization nor Marcus theory itself remain valid.

In the first part of this chapter we focus on rubrene, present its crystal structure, transfer integral distributions and the corresponding charge transport simulation results. We then study indolocarbazole followed by the two BBBT derivatives. In each case, differences to rubrene will be highlighted. Thereafter, details on the charge transport simulations for rate-based and semiclassical dynamics are given and their results are summarized and compared. Conclusions are drawn with respect to dimensionality and advice is given with regard to the ideal morphology of molecular crystals.

7. Organic Crystals

7.1. Rubrene

With a mobility of up to $\mu = 15\,\text{cm}^2/\text{Vs}$ in OFETs, rubrene has one of the highest mobilities among organic semiconductors and is hence one of the experimentally most studied materials [137, 134, 136]. Experimental efforts in creating nearly defect-free layers accomplished by vapor deposition using molecular beam epitaxy [132] as well as in purifying the material to prevent deep traps [140] led to substantial improvements of the mobility in rubrene-based devices over time. However, such improvements are costly and time-intensive. Hence it would be helpful to theoretically understand which materials are promising for such treatment based on their chemical structure and morphology alone.

Theoretically, charge transport in single crystals such as rubrene has often been considered band-like, due to their well-ordered structures and a negative temperature dependence of the mobility [141, 142]. Recently it was shown that the dynamic disorder in the transfer integrals at ambient conditions is of the same order as the transfer integrals themselves and therefore the electronic wave function cannot be delocalized over the entire crystal [74]. Rate-based simulations are likely to underestimate the mobility, because the average coupling along rubrene's main transport direction is of the same order of magnitude as its reorganization energy and thus neither the assumption of charge localization nor Marcus theory are valid. Nonetheless, we are interested in how the rate-based approach performs in this case. As is shown in a study using semi-classical dynamics, a treatment based on the diffusion limited by thermal disorder model is the most appropriate [75]. It not only takes into account that the charge carrier is spread over a limited number of neighboring molecules but also correctly reproduces the experimentally found negative temperature dependence of the mobility.

In this chapter we aim to understand why the chemical and crystal structure of rubrene, shown in fig. 7.1, render it such a highly conductive organic compound. To this end, connections between the morphology and the transfer integral distributions are illustrated and the mobility is calculated by both semi-classical dynamics and a rate-based approach. Finally, results are compared to two other molecular crystals in sections 7.4 and 7.5.

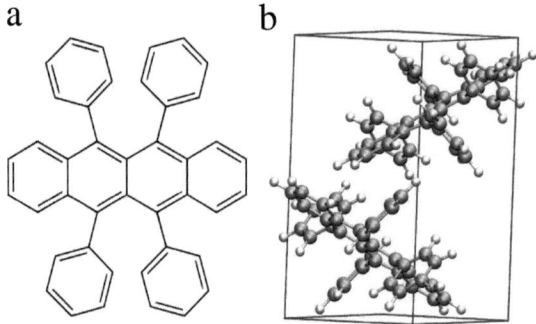

Fig. 7.1.: (a) Chemical structure of rubrene. (b) Unit cell of the corresponding single crystal.

7.1.1. Molecular dynamics simulations and charge transport parameters

MD simulations were performed using *Tinker* [143] in combination with the MM3 force field, exactly as it was done in ref. [75]. The system size was increased from 4 by 4 by 4 rubrene molecules to 8 by 8 by 4 to reduce finite size effects in the xy-plane, where charge transport occurs. The total of 17,920 atoms in the simulation box approached the limit of *Tinker*'s abilities and significantly larger box sizes could not be studied with the MM3 force field. The system was first equilibrated for 100 ps in an NPT ensemble, followed by a production run of 24 ps in an NVT ensemble, saving snapshots at 20 fs intervals. The short equilibration run was sufficient since the initial box was a 2 by 2 supercell of a very well equilibrated Rubrene system taken from ref. [75].

The reorganization energy was calculated based on a geometry optimized with B3LYP/6-311G(d,p) for a single molecule in vacuum. For a rubrene cation we obtained $\Lambda = 0.159$ eV. In rubrene the side chain phenyl rings contribute substantially to the HOMO orbital (see inset of fig. 7.2b) and mapping them in a rigid fashion yields transfer integral distributions that are in disagreement with those presented by Troisi in ref. [75], where the transfer integrals are obtained by directly calculating the corresponding Fock matrix elements built from the unperturbed density matrix [74]. To account for this in our simulations, we split the rubrene molecule into five different mapping sites, one for the core and one for each of the four phenyl side groups. Hence we used a linear combination of atomic ZINDO orbitals and rotated the orbitals corresponding to the phenyl groups separately from those of the core. Construction of a diabatic state then reproduced the transfer integral distribution presented in ref. [75].

7.1.2. Linking structure and transfer integral distributions

As already illustrated for BTBT (sec. 6.2), in crystals it is possible to relate the extrema in the transfer integral distributions to the corresponding neighbors on the crystal lattice. In addition, a comparison of the transfer integral distributions of the entire system for the main charge transport direction at a given time with the transfer integral distributions over time $J(t)$ for different neighboring pairs along the same direction reveals how strongly static and dynamic disorder influence the system.

For rubrene charge transport can occur only in the xy-plane, because along the z-direction centers of mass of the molecules are diagonally displaced very far, by approx. (0.38, 0.03, 1.38) nm, and the distance of closest approach is between the hydrogens of the side chain phenyls, which leads to negligible electronic coupling. There are three different major directions of electronic coupling, displayed in fig. 7.2a. A cut-off radius of $r_c = 1.2$ nm was chosen for the neighbor search to take into account exactly six neighbors per molecule, i.e. two for each transport direction. Along the A direction the neighbors are cofacially aligned, though shifted by on average 0.714 nm with respect to each other along the x-axis. This shift is the distance between the centers of mass and thus the distance a charge hops upon transfer in the KMC simulations, which is a lot further than the distance of closest approach. The average coupling for direction A (calculated for a single snapshot) shown in fig. 7.2b is very high, but has a broad distribution $J_A = 0.078 \pm 0.030$ eV. In directions B and C neighboring molecules are tilted with respect to each other, so that there is no cofacial alignment between them as for direction A and, consequently, the electronic coupling is lower. This is evidenced by the respective transfer integral distributions shown in fig. 7.2b. Both exhibit a very pronounced, sharp peak at $J_B = J_C = -0.010 \pm 0.006$. For all directions the case of zero coupling between neighbors is at the very edge of the distributions and thus charge transport is never truly disrupted in the xy-plane.

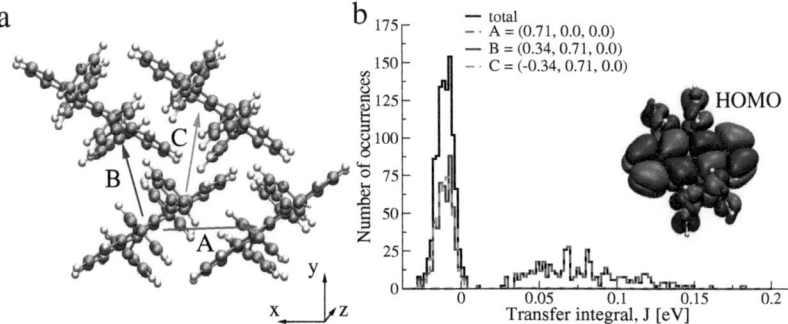

Fig. 7.2.: (a) Nearest neighbor pairs of the rubrene crystal in directions A, B and C corresponding to the three main transport directions in the xy plane. (b) shows the respective transfer integral distributions of the entire system (black solid line) and those belonging to directions A, B and C.

To show whether the width of the transfer integral distribution in A direction is due to static or dynamic disorder, we analyzed the $J(t)$ distributions for neighboring pairs of type A based on 1000 snapshots taken every 20 fs. A comparison of temporal transfer integral distributions for five neighboring pairs with a configurational distribution based on all neighbors of type A in a single snapshot, rescaled for ease of comparison, is presented in fig. 7.3a. The width of each of the temporal distributions is similar to that of the configurational distribution. This means that thermal fluctuations between neighbors lead to variations in transfer integrals which are of the same magnitude as variations throughout the system. Thus thermal fluctuations and dynamic disorder alone suffice to explain the disorder in transfer integrals.

A connectivity graph for the xy plane of rubrene is shown in fig. 7.3b. Here grey spheres represent the centers of mass of the molecules, thickness of the bonds between them corresponds to the magnitude of the connecting transfer integral and color to the sign, with red being negative and blue positive. No bonds are drawn in case the transfer integral between neighbors is below 0.005 eV. It is no surprise that the connectivity graph shows no defects throughout. Also the sign of the transfer integrals is always identical in a given direction, i.e. molecular vibrations are not strong enough to lead to a complete loss of coupling between neighboring molecules.

7.1.3. Charge transport simulations

In the rate-based simulations, rubrene shows a high mobility of $8.14\,\text{cm}^2/\text{Vs}$ in the x direction, i.e. along direction A. Perpendicular, in y direction, the mobility is $0.45\,\text{cm}^2/\text{Vs}$ while it is only $3 \cdot 10^{-8}\,\text{cm}^2/Vs$ in the z direction. Rubrene is thus a 2D transporting compound with the highest mobility along the direction of shifted cofacially aligned molecules. However, the rate-based mobility even along the main axis (x) is below that experimentally measured in

7.2 Indolocarbazole

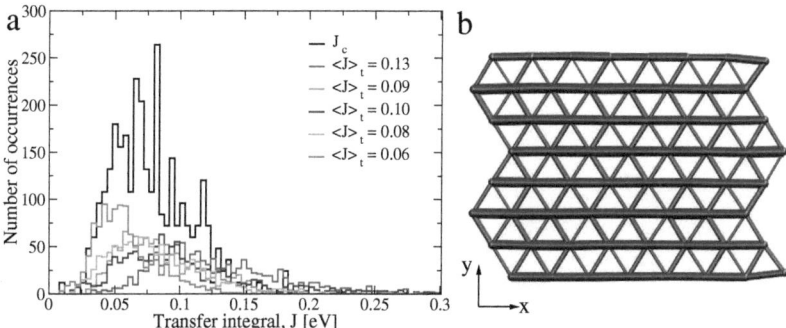

Fig. 7.3.: (a) Distribution of transfer integrals in rubrene with respect to configuration (J_c) compared to that due to time fluctuations for different neighboring pairs of type A. Thermal fluctuations alone suffice to explain the large standard deviation of the configuration transfer integral distribution. (b) Corresponding connectivity graph. Grey spheres represent the centers of mass, thickness of bonds is proportional to the connecting transfer integral, colors represent the sign, i.e. red is negative and blue positive. No bonds are drawn if the transfer integral between neighbors is below 0.005 eV.

OFETs (15 cm^2/Vs [134, 135, 136, 137]), but still of the same order of magnitude. A rate-based treatment is expected to underestimate the experimental values, since hopping transport is slower than small polaron transport in well-ordered, strongly coupled systems. Also, a rate-based description is not strictly valid in case of rubrene, because the transfer integrals in A direction ($J_A = 0.078$ eV) are of the same order as the reorganization energy ($\Lambda = 0.159$ eV).

The appropriate way to treat rubrene is in the small polaron regime with the diffusion limited by thermal disorder model using semi-classical dynamics. Such simulations predict a mobility of 64 cm^2/Vs for rubrene, which is significantly higher than the rate-based value. It also exceeds the experimental OFET mobility, but is of the same order of magnitude. Since static disorder is not significant in rubrene and there are no low rates in the A direction, we believe that the one-dimensionality of SCD does not neglect important system properties and thus SCD results are the upper limit of the mobility, which may be achieved in a perfect crystal.

Details on both rate-based and SCD simulations as well as an in-depth comparison to the other compounds follows in sections 7.4 and 7.5.

7.2. Indolocarbazole

The first compound we compare to rubrene is indolo[2,3-b]carbazole. The conjugated core of indolocarbazole is similar to pentacene and the presence of a nitrogen in the core introduces a binding site, which may be used for the attachment of functional groups. Functional groups can be used to vary the solubility and tune the molecular arrangement to favor charge transport.

A variety of indolocarbazole derivatives were synthesized for application in OLEDs within the international research training group program by Norma Wrobel [138]. We focus here on the most basic indolo[2,3-b]carbazole compound with simple CH$_3$ side groups. The chemical structure of the single molecule (a) and the corresponding unit cell of the crystal (b) according to x-ray measurements [138] are shown in fig. 7.4. The single crystal unit cell has P2$_1$/c symmetry with unit cell vectors of length $a = 1.1229$, $b = 0.78561$ and $c = 0.9668$ nm and corresponding angles $\alpha = \gamma = 90°$ and $\beta = 94.79°$.

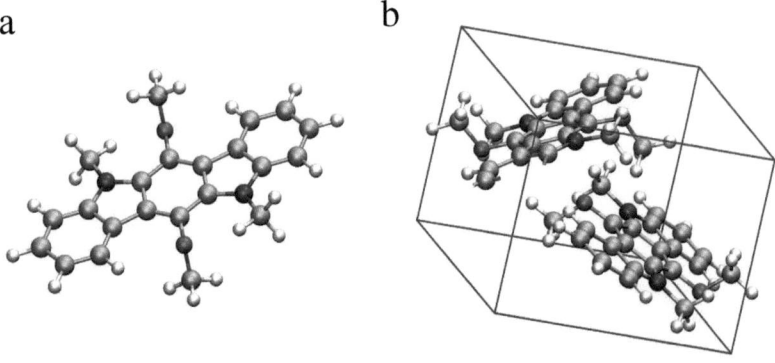

Fig. 7.4.: (a) Chemical structure of the indolocarbazole compound with CH_3 side chains. (b) Unit cell of the corresponding single crystal.

7.2.1. Molecular dynamics simulations and charge transport parameters

The force field required for MD simulations is based on that created for carbazole (sec. 5.1), but in this case OPLS all atom parameters were used to describe the CH_3 side chains. Details may be found in the appendix A.3.2. All MD simulations were carried out using a 12 by 12 by 12 unit cell sized box containing a total of 3456 molecules or 158,976 atoms. At first the system was equilibrated in a 350 ps NPT run using Berendsen thermostat and barostat. Initial velocities were randomly chosen from a Boltzmann distribution at 300 K. This was followed by a 1 ns NVT simulation using the Nosé Hoover thermostat. After ensuring that energy, temperature and pressure had converged, production runs were performed for 20 ps with snapshots taken every 20 fs. Initially, we used the Nosé Hoover thermostat, but the time constant of the thermostat can be seen in the temperature fluctuations of the system on such timescales and thus introduces spurious correlations in the transfer integral correlations. This effect is known to occur in highly harmonic crystals with small unit cells, whose phonon modes are coupled to the thermodynamics of the system and thus the thermostat. To avoid this effect, the production runs use the NVE ensemble. To prevent energy loss during the NVE simulation it is important to update the neighbor list at every step and not every five steps as is usually the case in MD simulations. We confirmed that the NVE runs were sufficiently equilibrated by analyzing the velocity distribution of the system and confirming that velocities of both carbons and hydrogens obey a Boltzmann distribution [1].

---

[1] We also compared the NVE results with NVT using the velocity rescaling thermostat, which was implemented in *Gromacs* after the completion of the indolocarbazole study, and found them to be identical on such short time

7.2 Indolocarbazole

The reorganization energy was calculated based on a single molecule in vacuum optimized with B3LYP using the 6-311G(d,p) basis set. We obtained $\Lambda = 0.212$ eV, which is about thirty percent higher than that of rubrene. The highest occupied molecular orbitals were computed based on the optimized geometries, i.e. for the entire indolocarbazole molecule. The shape of the HOMO is illustrated in the inset of fig. 7.5a.

7.2.2. Linking structure and transfer integral distributions

Charge transport in indolocarbazole occurs between three different types of neighboring pairs in the crystal, which are labeled A, B and C in fig. 7.5b. Neighbors of type A and B lie in the yz-plane with the centers of mass of the molecules separated by vectors $(0.053, 0.423, -0.476)$ and $(0.053, -0.423, -0.476)$ nm, respectively. The transfer integral distributions for A and B in a single snapshot are shown by the red and blue curves in fig. 7.5a and are found to be similar, with $J = 0.010 \pm 0.003$ eV based on averaging over respective pairs in a single snapshot. Neighbors of type C are connected by the vector $(1.111, 0.0, 0.0)$. However, as can be seen from the bottom of fig. 7.5b, the minimal distance between them is much smaller and enables transport along the x direction of the crystal. The corresponding transfer integral distribution is centered at $J = -0.006 \pm 0.003$ eV. To ensure that all neighbors of type C are taken into account, a cut-off radius of 1.2 nm was chosen for the neighbor search. The large additional peak at zero in the transfer integral distribution (see fig. 7.5a) is an artifact due to second-order neighbors, which do not influence the simulation results, because neighbors with low coupling are removed prior to the KMC runs.

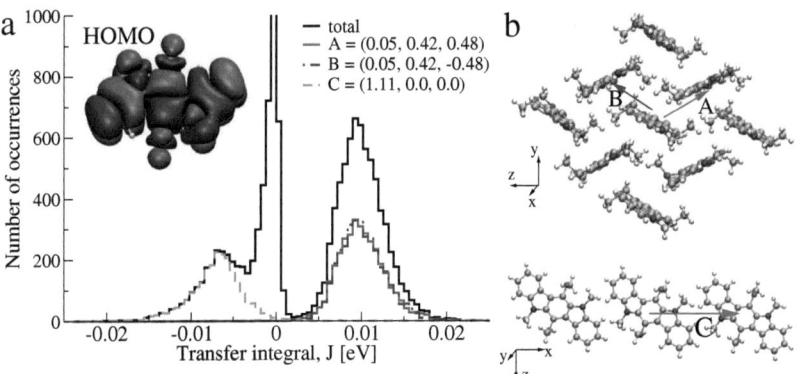

Fig. 7.5.: (a) Transfer integral distribution for a single snapshot of an indolocarbazole crystal subdivided into three main transport directions. (b) Neighboring pairs corresponding to the transport directions.

To identify the source of the variation of the transfer integrals, in fig. 7.6a we compare the transfer integral distribution of a single snapshot for the A direction with the temporal distribution $J(t)$ of three exemplary neighboring pairs of type A. The $J(t)$ distributions are based on 1000

---

scales. Thus results obtained here are thus perfectly comparable with NVT results using velocity averaging for the other crystalline systems in this chapter.

frames 20 fs apart. Out of these three pairs the one indicated by the red line in fig. 7.6a has the lowest average transfer integral amounting to ($< J(t) > = 0.0068$ eV). The corresponding $J(t)$ distribution is narrow and located close to the lower tail of the configurational distribution J_{conf}. In contrast, the other two pairs have average transfer integrals of 0.0086 and 0.0126 eV, close to the center of the full distribution. In addition, their distributions are also very broad with a width similar to that of J_{conf}. This indicates that static disorder in the system is small and transfer integral fluctuations are mainly due to thermal noise. Already, from the three distinct neighbor connections, it can be seen that indolocarbazole features a percolation network in three dimensions, in contrast to the 2D connectivity of rubrene.

Connectivity graphs of different planes in the system using the same cut-off as for rubrene, i.e. $J = 0.005$ eV, are shown on the right of fig. 7.6. Since the transfer integrals for indolocarbazole are are about an order of magnitude smaller than in rubrene's main transport direction, the bonds drawn in fig. 7.6b and c are slimmer. This difference is explained by the CH_3 and O-CH_3 side chains, which prevent a close approach of the conjugated core of neighboring molecules. However, the resulting absolute values of the transfer integrals are of the same order as those obtained for the weakly coupling B and C directions in rubrene. The transfer integral distributions alone do not reveal a preferred transport direction, although electronic coupling in C and thus x-direction is almost a factor of two lower than in A and B direction. Consequently, the connectivity network in fig. 7.6b shows only very few defects and hence plenty percolation pathways in the yz-plane. The respective graph for the xz-plane shows a number of disruptions in connectivity due to the lower coupling in C direction (and the high chosen cut-off of $J = 0.005$ eV), but does not render charge transport unlikely. Indolocarbazole is thus an isotropic 3D transporting compound. Combined with the little static disorder in the system, this results in a well-connected charge carrier percolation network spanning all three spatial dimensions.

7.2.3. Charge transport simulations

Rate-based simulations yield mobilities of approximately 0.09 cm^2/Vs in both x and y direction and 0.06 cm^2/Vs in z-direction. As one would expect from the similar transfer integral distributions, charge transport is equally good in all directions. It is almost two orders of magnitude below that of rubrene in its main transport direction, but less than one order of magnitude below rubrene's y-direction. This is surprisingly good when taking into account the far weaker transfer integrals and the higher reorganization energy and can only be explained by the almost defect-free 3D charge percolation network. SCD simulations predict a mobility of 1.7 cm^2/Vs, which is again more than an order of magnitude below that of rubrene, but still surprisingly high, since the isotropic percolation network does not positively influence the one-dimensional SCD simulations. The narrow width of the transfer integral distributions may be responsible for this.

Comparing the two approaches, since the average coupling $J \approx 0.01$ eV for all directions is significantly smaller than the reorganization energy $\Lambda = 0.212$ eV, a rate-based treatment of charge transport is justified despite the low static disorder. However, for an indolocarbazole compound with side chains allowing a closer approach between the conjugated cores of the molecules, this is likely to change.

7.2 Indolocarbazole

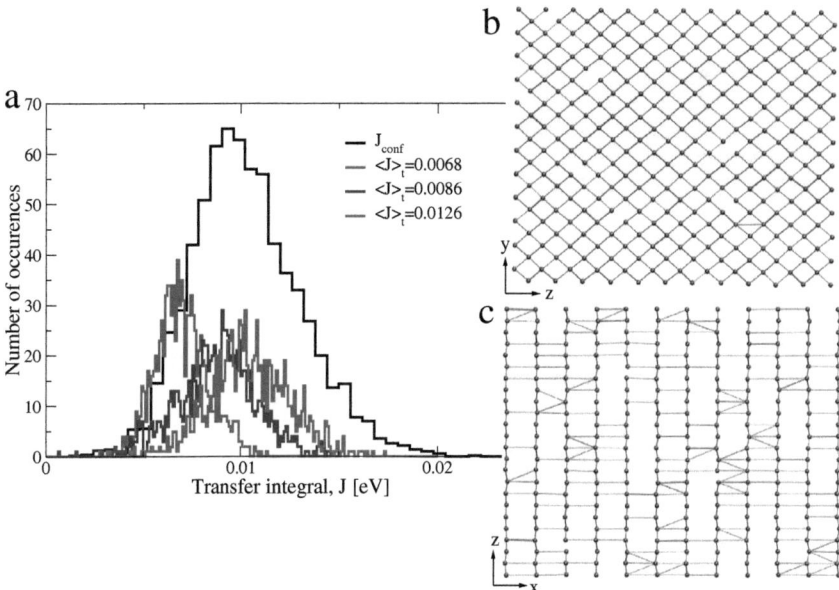

Fig. 7.6.: (a) Transfer integral distribution of indolocarbazole for a single snapshot J_{conf} compared to the $J(t)$ distribution for three neighboring pairs of type A. Most of the disorder may be explained by thermal fluctuations. Connectivity graphs for the (b) yz and (c) xy-plane of the indolocarbazole crystal. Grey spheres represent centers of mass, bond width is proportional to magnitude of transfer integral, no bonds are drawn if $J < 0.005$ eV.

7.3. Benzo[1,2-b:4,5-b']bis[b]benzothiophene (BBBT)

BBBT is a conjugated molecule with a rigid fused-ring structure similar to pentacene and BTBT. While pentacene shows one of the highest mobilities among organic π-conjugated materials in OFETs, devices built from it suffer from low stability and short lifetimes under ambient conditions. Materials based on fused thiophene aromatics have been shown to have high stability paired with good OFET characteristics. BBBT is an analog of pentacene with a rigid, linearly conjugated structure but a thiophene-based core. A straight-forward, high-yield synthesis of BBBT and its alkyl substituted derivatives has recently been developed [139].

In the same study, organic field effect transistors have been made by solution-processing using two different BBBT derivatives: one without and one with C_4H_9 side chains. The x-ray structures of the corresponding unit cells are displayed in fig. 7.7. The unit cell for the compound without side chains, labeled as BBBT(1) in the following, has P1 symmetry with box lengths $a = 0.949$, $b = 0.590$ and $c = 1.158$ nm and corresponding angles $\alpha = 90.0°, \beta = 102.94°$ and $\gamma = 90.0°$. It contains two molecules, which are tilted with respect to each other. The unit cell for the compound with C_4H_9 side chains, labeled by BBBT(2) from this point on, also has P1 symmetry with $a = 0.458$, $b = 9.223$ and $c = 1.249$ nm. The angles are $\alpha = 89.85°, \beta = 79.59°$ and $\gamma = 79.16°$. In contrast to BBBT(1) it contains only one molecule and the molecules come very close in almost perfect cofacial alignment in the crystal.

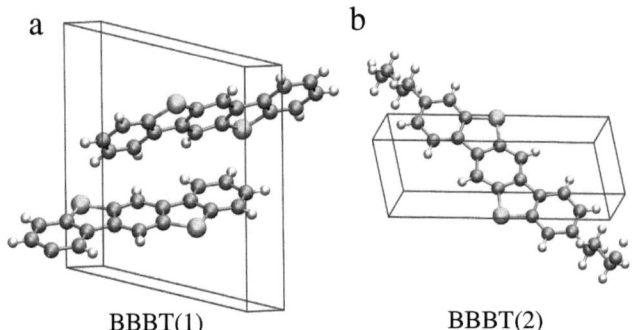

Fig. 7.7.: Unit cells of BBBT (a) without and (b) with C_4H_9 side chains.

7.3.1. Molecular dynamics simulations and charge transport parameters

The force field parameters used for the study of BBBT are almost identical to those of BTBT with changes to account for the presence of the central phenyl ring. Details are given in the appendix A.3.3. The MD runs for both compounds were performed using a box consisting of 12 by 12 by 12 unit cells, i.e. 3456 molecules for BBBT(1) and 1728 for BBBT(2) corresponding to 103,680 and 62,208 atoms. Initial velocities were chosen randomly from a Boltzmann distribution at 300 K. NPT equilibration runs lasted 2 ns with the velocity rescaling thermostat and Berendsen pressure coupling. They were followed by 1 ns NVT equilibration runs using the equilibrated box size and the velocity rescaling thermostat. NVT production runs using the velocity rescaling thermostat lasted 22 ps with snapshots taken every 20 fs.

The molecular orbitals and thus charge transport properties of BBBT are assumed not to be influenced by attachment of alkyl side chains. Thus the reorganization energy for both BBBT compounds is calculated from a single molecule without side chains in vacuum optimized using B3LYP/6-311G(d,p) and amounts to $\Lambda = 0.120\,\mathrm{eV}$. This value is even lower than that of rubrene and thus more favorable for charge transport. The HOMO was computed based on the same optimized geometry. It is illustrated in the inset of fig. 7.9a.

7.3.2. Linking structure and transfer integral distributions

BBBT without side chains (1)

In the crystal structure of BBBT(1) there are four types of neighboring pairs partaking in charge transport, which are indicated in fig. 7.8. The three types of pairs A, B and C contribute mainly to transport in the yz-plane. In particular, A has the lowest center of mass distance with a connecting vector of $(0.0, 0.587, 0.0)\,\mathrm{nm}$. The molecules are in strongly tilted cofacial alignment leading to an average transfer integral of $J_A = 0.011 \pm 0.004\,\mathrm{eV}$. Neighbors of type B and C are at identical distance with characteristic vectors $(0.355, 0.275, 0.582)$ and $(0.355, -0.275, 0.582)\,\mathrm{nm}$. Despite the proximity of the neighbors, as shown in fig. 7.8a, the tilt of the molecules in opposite directions with the cores almost perpendicular to each other allows for little overlap between orbitals, resulting in a weak electronic coupling of $J_B = J_C = 0.001 \pm 0.009\,\mathrm{eV}$. The best electronic coupling is found along the x axis in direction D, where neighbors are connected by the vector $(0.926, 0.0, 0.0)$, since the molecules are in shifted cofacial alignment as shown in fig. 7.8b. A cut-off radius of $r_c = 1.1\,\mathrm{nm}$ was chosen to take all neighbors of type D into account. The corresponding transfer integral distribution is strong but broad with $J_D = 0.036 \pm 0.020$. In fig. 7.9a the transfer integral distribution for the entire system and subdivided into the four aforementioned directions based on a single snapshot is shown. As for indolocarbazole, the peak at zero in the transfer integral distribution of the entire system is explained by second-order neighbors.

We also compared the distribution of transfer integrals in the main direction D for a single snapshot with the $J(t)$ distribution of four exemplary neighboring pairs over 1000 snapshots, taken every 20 fs. As can be seen in fig. 7.9b, the width of the $J(t)$ distributions is similar to that of the configurational distribution J_{conf}. This indicates that, as for rubrene and indolocarbazole, static disorder is not significant, and the width of the disorder in transfer integrals can be explained by thermal fluctuations alone. The connectivity graph, shown in fig. 7.12a in comparison to that of BBBT(2), drawn for neighbors with coupling above $J = 0.005\,\mathrm{eV}$ gives a clear picture of strong transport along the x-axis with weak interconnections in the yz-plane. Therefore, as one might expect due to the lack of side chains, BBBT(1) is a three-dimensional semiconductor. In contrast to the also 3D conducting indolocarbazole, BBBT(1) possesses a strongly preferred transport direction.

BBBT with C4 side chains (2)

Attachment of alkyl side chains changes the crystal structure of BBBT(1) into one with a significantly different neighbor order. BBBT(2) aligns in a stacked columnar phase, where each

7. Organic Crystals

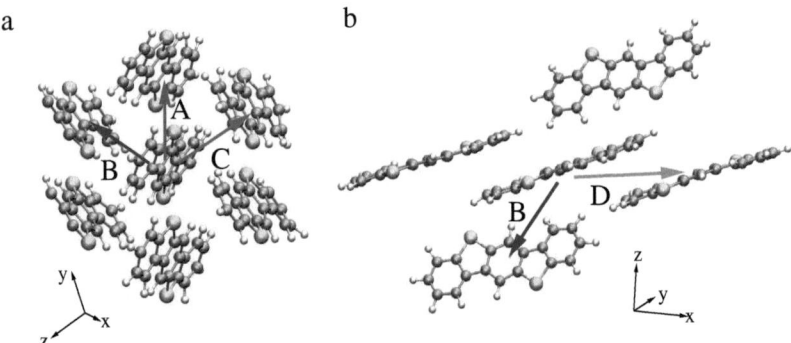

Fig. 7.8.: Nearest neighbor pairs contributing to charge transport in BBBT(1) shown in the (a) yz and (b) xz plane.

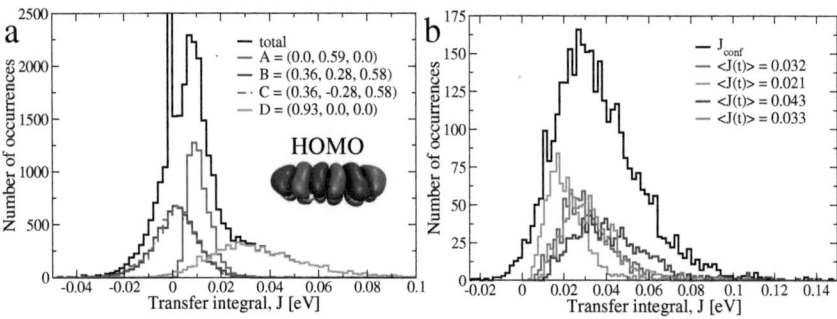

Fig. 7.9.: (a) Transfer integral distribution of the entire system for one snapshot subdivided into the four directions A, B, C and D. Corresponding neighboring pairs are shown in fig. 7.8. The HOMO used in the transfer integral calculations is depicted in the inset.(b) Comparison of $J(t)$ distributions for four different pairs of type D with the transfer integral distribution J_{conf} for type D neighbors in an entire snapshot.

molecule has only two close neighbors. The molecules have practically perfect cofacial alignment along the column, i.e. in $A = (0.47, -0.01, 0.01)$ nm direction as illustrated in fig. 7.10a. The interplanar separation of this lamellar structure is only 0.342 nm, which leads to a strong electronic coupling of $J_A = -0.130 \pm 0.042$ eV. Notably, this value exceeds even the one along rubrene's main transport direction. However, unlike in the case of rubrene, no transport in other directions can be expected due to the presence of the side chains and the alignment of the molecules in isolating neighboring stacks. As shown in fig. 7.10b, neighbors perpendicular to the stacking direction can only interact via the side hydrogens resulting in low orbital overlap. Since maximum values for such coupling are on the order of 10^{-3} eV, these directions cannot be completely neglected and consequently a cut-off radius of 1 nm was chosen to take into account a few neighbors off of the main transport direction A. The attachment of the side chains thus changes the crystalline ordering in such a way that BBBT(2) becomes a one-dimensional conductor, with transport only along the stacking direction A of the columns. This is supported by the transfer integral distribution within the system, shown in fig. 7.11, which splits into two parts: the distribution due to neighbors along the A direction and a peak at zero, which contains the contributions of all neighbors in directions other than A as well as second order A neighbors, which are responsible for the two small maxima beside the central zero peak.

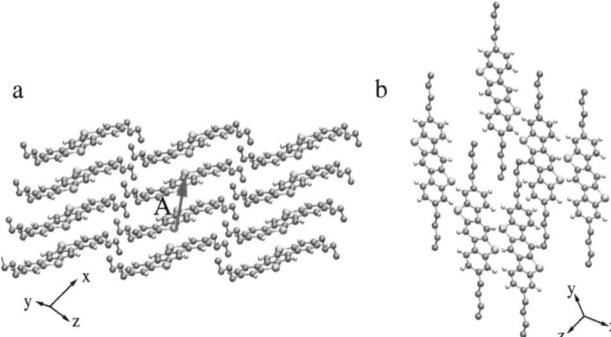

Fig. 7.10.: (a) 1D cofacial stacking in the A direction responsible for the strong coupling along the column in BBBT(2). (b) Shift and separation of stacks are responsible for bad coupling in all other directions.

To study the source of the disorder in the transfer integrals, we restrict our discussion to neighbors along the columnar stacking direction A. In fig. 7.11b we compare the $J(t)$ distributions of seven exemplary neighbor pairs of type A with the corresponding configurational distribution J_{conf}. In contrast to BBBT(1), the width of no temporal transfer integral distribution rivals that of the configurational one. This indicates the presence of significant static disorder along the column, which in analogy to BTBT (sec. 6.2) may be explained by the slow motions of the soft side chains and resulting displacements of the molecules in direction of their long axis. In contrast to BTBT, static disorder is not strong enough to deter the electronic coupling between neighboring molecules in the A direction and prevent charge transport. However, we expect the reduction in mobility due to static disorder to become more significant in longer MD simulations of larger systems and thus also in OFET measurements, since the lowest transfer integral

7. Organic Crystals

value along the column limits the mobility.

Not surprisingly, the connectivity graph shown in fig. 7.12b, which displays bonds only for neighbors with transfer integrals $J > 0.005$ eV, shows connectivity only along the x-direction. As indicated by the width of the bonds in the connectivity graph, the electronic coupling between most neighbors is very strong, but for some it is significantly weaker. Note that in comparison to BBBT(1) (see fig. 7.12a), electronic coupling in x-direction becomes stronger on average, but the three-dimensionality and with it the existence of percolation paths to circumvent defects along the main direction of coupling are lost.

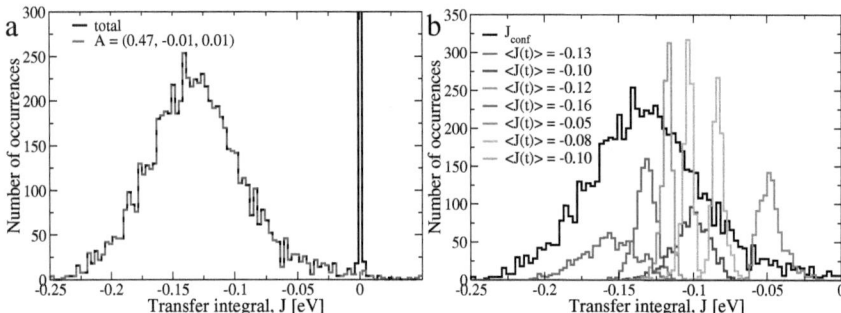

Fig. 7.11.: (a) Distribution of BBBT(2) transfer integrals in an entire snapshot along with the distribution in A direction. The peak at zero is cut for clarity. (b) $J(t)$ distributions for seven different type A pairs compared to the configurational distribution $\bar{J}_c$.

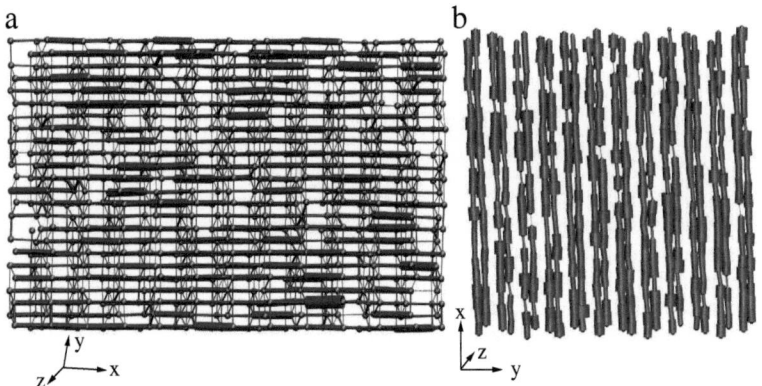

Fig. 7.12.: Connectivity graph for BBBT (a) without and (b) with C_4H_9 side chains. Grey spheres represent centers of mass, bond width is proportional to coupling between neighbors, no bonds are shown if $J < 0.005$ eV. Red indicates negative and blue positive transfer integral values.

7.3.3. Charge transport simulations

Kinetic Monte Carlo simulations based on Marcus rates predict a mobility of $4.5\,\text{cm}^2/\text{Vs}$ in $x-$, 0.33 in $y-$ and 0.29 in z-direction for BBBT(1). It is therefore the most similar to rubrene, especially since molecules are also in shifted cofacial alignment along its main transport direction. While the mobility is predicted to be slightly weaker in both x and y direction, BBBT(1) possesses mobility in the z direction as well, which is of the same order as its own and rubrene's y-direction. It is therefore a truly 3D transporting system. When comparing charge transport in an ideal crystal to that in the equilibrated system, the charge mobility along the x-axis increases by a factor of three to $13.2\,\text{cm}^2/\text{Vs}$, but remains almost the same along y- and z-axes. Thus the main coupling direction is most sensitive to introduction of disorder.

For BBBT(2), rate-based simulations predict a mobility of $12.5\,\text{cm}^2/\text{Vs}$ along the x-axis, which exceeds even that of rubrene. This is not surprising, since the mean of the distribution of transfer integrals in that direction is higher than that of rubrene and the reorganization energy of BBBT is lower. However, transport along all other directions is negligible, i.e. $\approx 10^{-4}$ and $10^{-8}\,\text{cm}^2/\text{Vs}$ in y- and z-direction, respectively. BBBT(2) is thus a 1D transporting compound. Even in the ideal crystal, no transport is seen in y- and z-direction, while transport in x direction triples to $36.4\,\text{cm}^2/\text{Vs}$. It is important to keep in mind that for BBBT(2) the mobility will also depend on the tail of the transfer integral distributions as is emphasized in the chapters on carbazole and hexabenzocoronene. Thus, a single low transfer integral will reduce charge transport in a column. This means that the mobility in the BBBT(2) simulation depends on the starting position of the charge carrier since it is confined to the same column for the remainder of the simulation. Indeed, in small one-dimensional systems the width of the mobility distribution due to different columns is expected to decrease proportional to $1/\sqrt{N}$. In our simulations with twelve molecules per column, we thus observe variations by up to a factor of 3.5 in the mobility depending on the starting postion. In contrast, the starting position only causes mobility variations of a few percent for the other systems.

When charge transport is treated using semi-classical dynamics, the mobility along the main transport directions of BBBT(1) and BBBT(2) is predicted to be 9.43 and $8.15\,\text{cm}^2/\text{Vs}$, respectively. Thus that of BBBT(1) actually exceeds that of BBBT(2) despite the excellent cofacial coupling. This is explained by the fact that the lower coupling in BBBT(1) is partially compensated by the larger distance of 0.93 nm a charge travels between neighboring molecules in the shifted cofacial direction of BBBT(1) while it travels only 0.47 nm in BBBT(2). In addition, the Peierls coupling constant for BBBT(1) is also only a quarter of that of BBBT(2), since the molecules have lower weight and the transfer integral distributions are more narrow.

For the two BBBT compounds the coupling in the main direction is again of the same order as the reorganization energy, indicating that semi-classical dynamics simulations are the more realistic description. This is supported by the fact that static disorder has no major influence in BBBT(1), while BBBT(2) is actually a one-dimensional conductor.

7.4. Charge transport using rate equations

Kinetic Monte Carlo simulations based on Marcus rates were performed using time of flight, diffusion and velocity averaging analysis (see sec. 4.4.4), all of which were in agreement with each other. Here only results obtained with velocity averaging are presented. The dipole moment of each molecule was calculated for the optimized geometries of a single molecule in vacuum using B3LYP/6-311G(d,p). The low results for all compounds, namely 0.0474 Debye for rubrene, 0.0009 for indolocarbazole and 0.0006 for BBBT, led us to ignore energetic disorder arising from electrostatic interactions. Thus the free energy difference between molecules was set to $\Delta G = e\mathbf{E}\mathbf{r}_{ij}$, where e is the elementary charge, $\mathbf{r}_{ij}$ is the vector connecting molecules i and j and E is the externally applied electric field chosen to have a magnitude of 10^7 V/m throughout. The field vector was always aligned in parallel to the direction for which transport is calculated. There is also no diagonal disorder, since we assume that all molecules in the system are identical and thus have the same HOMO energy level. The results in the three main directions x, y and z are based on velocity averaging runs averaged over multiple snapshots and different starting positions therein. Different directions are analyzed by rotating the external field along the desired direction. To emphasize the influence of disorder, the above simulations are compared with those done on the ideal crystal configuration obtained by multiplication of the x-ray unit cell without further NPT equilibration.

The velocity averaging results for all crystalline compounds are based on 10^{-8} second long runs for 500 frames at 10 different starting positions each. Results in the diagonal directions are based on 100 frames with 10 starting positions each. For an ideal crystal only one frame is used and averaging is done over 10^{-7} sec runs and 100 different starting positions. The resulting mobilities (cm^2/Vs) are listed in tab. 7.1.

Compound	x	y	z	FET (exp.)
Indolocarbazole	0.092	0.063	0.090	
Rubrene	8.142	0.450	$3 \cdot 10^{-8}$	15 [134, 135, 136, 137]
BBBT (1)	4.529	0.328	0.294	10^{-2} [139]
BBBT (2)	12.51	10^{-4}	$4 \cdot 10^{-8}$	10^{-3} [139]
Compound	x(i)	y(i)	z(i)	xyz(i)
Indolocarbazole	0.099	0.060	0.083	0.076
BBBT (1)	13.195	0.324	0.394	4.801
BBBT (2)	36.39	$4 \cdot 10^{-6}$	10^{-5}	12.13

Tab. 7.1.: Mobilities in cm^2/Vs for Indolocarbazole, Rubrene and BBBT without and with C$_4$H$_9$ side chains. Experimental field-effect transistor data is given for Rubrene and BBBT, no direction of transport was specified in the corresponding publications. (i) refers to the mobility measured for the ideal crystal created based on the x-ray unit cell.

Rubrene and the BBBT compounds all show very high mobilities along the directions of their strongest coupling in the simulations, i.e. along the A direction in case of rubrene and BBBT(2) and along D in BBBT(1), where neighboring molecules are either in cofacial or shifted cofacial alignment. The mobility in indolocarbazole is almost two orders of magnitude lower, but with equally good transport in all three dimensions. BBBT(1) is also 3D transporting. However, the y and z directions are an order of magnitude weaker than the x direction. Rubrene is a 2D transporting material, also with an order of magnitude stronger transport along the x than along the y axis. BBBT(2) shows no transport perpendicular to the main direction. In the ideal crystal, the mobility of indolocarbazole is almost identical to that in the equilibrated system. In case of BBBT the transport in the main direction triples, while the perpendicular directions remain unchanged. Experimentally, rubrene gives high mobilities on the same order as in the simulations, while both BBBT compounds show three or four orders of magnitude lower transport in OFET measurements. Reasons will be given below.

The preferred transport directions in the crystals can be visualized by diffusion pathways. Without an applied electric field the charge drifts randomly along the preferred transport directions. The shape of the resulting random walk therefore corresponds to the mobility tensor in the limit of large times. In case of rubrene the charge is confined to the crystal plane in which it is generated while for BBBT(1) the charge can diffuse in three-dimensions with a clear preference on the x-direction. The corresponding diffusion path is shown in fig. 7.13. In case of BBBT(2) diffusion paths collapse to straight lines, since the charge travels almost solely along the x-direction, i.e. it travels at least 100 times further in x than in y or z direction in any given time interval.

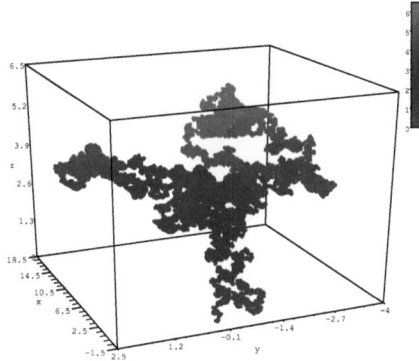

Fig. 7.13.: Three-dimensional diffusion pathway BBBT(1). For clarity the color scale denotes displacement along the z-axis. Note the different axes scalings.

7.4.1. Cluster analysis

To ease the comparison between the charge transport properties of the four different crystalline structures, it is helpful to investigate the connectivity of the system depending on the transfer integral between neighbors, i.e. by performing a cluster analysis which will result in effective transfer integrals for all systems. Here, all molecules which are connected by pathways in which all transfer integrals J exceed a certain value are considered to belong to a cluster. The largest cluster size with respect to transfer integral cut-off for all four compounds is shown in fig. 7.14. The size of the clusters is given in percent of the system size. For the three-dimensional conductors the maximum cluster size at a cut-off of 10^{-4} eV is of course the entire system. Since Rubrene only shows two-dimensional connectivity and contains four planes in the simulation box, the maximum cluster size is an entire plane and thus 25% of the system. For BBBT(2) the coupling perpendicular to the main axis is so low that the cluster size immediately collapses to one-dimensional arrays. Since the box consists of 12 by 12 by 12 unit cells, this corresponds to $1/(12 \cdot 12) = 0.68\%$ of the system. A closer look at the curve reveals, that the one-dimensional clusters begin to break down at 0.1 eV. The analysis clearly outlines why BBBT(1) has a higher mobility than Indolocarbazole, because the effective coupling is a factor of two lower in the latter case. For Rubrene, the cluster size diminishes stepwise starting at a later point and finally drops to a second plateau corresponding to one-dimensional clusters $(1/(4 \cdot 8) = 3.125\%)$ along the main coupling direction at 0.16 eV.

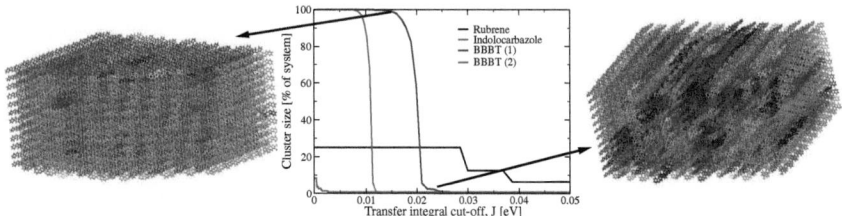

Fig. 7.14.: Size of largest cluster with respect to transfer integral cut-off. The cluster distributions are shown for BBBT(1) for cut-offs of $J = 0.018$ and 0.021 eV. Each cluster is represented by a different random color.

7.4.2. Comparison of experimental and theoretical results

Simulations predict that rubrene has the highest mobility followed by the two BBBT compounds and finally indolocarbazole with the lowest mobility. This relative order is in agreement with the experimental evidence. However, when comparing the absolute values obtained for the charge mobility by simulations to OFET data the agreement is good for rubrene (OFET: 15, simulation: $8 \, \text{cm}^2/\text{Vs}$), but the values differ by three to four orders of magnitude between experimental measurements and simulations for BBBT(1) and (2) (OFET: 10^{-2}, 10^{-3}, simulation: 4.5, $12.5 \, \text{cm}^2/\text{Vs}$) [139, 134, 135, 136, 137]. Of course, we cannot expect to exactly reproduce experimental results due to various approximations made in the simulations. The small size of the simulated systems and the short simulation times do not allow grain boundaries or severe defects to arise when running simulations based on an ideal x-ray structure. Also, traps due to impurities in the materials are neglected, since all simulated molecules are chemically identical. In the influence of interfaces between different layers in OFETs is not treated, because we deal with bulk systems only. Energetic disorder due to different HOMO energies which may arise from fluctuations in molecular geometry as well as due to electrostatic interactions is also neglected. All above mentioned effects would reduce the mobility in simulations. Nonetheless, agreement with OFET measurements for rubrene is very good and there are some experimental reasons why the measured mobility for rubrene is so much higher than for BBBT. First, rubrene single crystal layers were built by vapor deposition while BBBT was deposited by spin coating. Vapor deposition allows controlled growth of crystal layers by adjusting growth parameters such as deposition temperature, source temperature and growth time [144]. For example molecular beam epitaxy [132], where atoms in the evaporated beam do not interact with each other until they reach the deposition surface, allows each molecule to perfectly align within the forming layer. Therefore, disorder in rubrene layers is expected to be substantially less than in BBBT. Second, synthesis of rubrene has been improved over time and the purity of the layers is extremely high [140]. BBBT on the other hand has only recently been synthesized, hence layers are less pure, thereby introducing a number of molecules with different chemical structure. This may lead to significant diagonal disorder due to different HOMO levels and introduce deep traps in the system. Neither effect is included in our simulations. Third, BBBT(2) is a one-dimensional conductor with very strong electronic coupling. On the timescale of our simulations no significant static disorder or defects can be seen, but in a real spin cast system we would expect a number of defects, each of which could potentially prevent transport along a column. Hence it is not surprising, that in experiment the 1D conducting BBBT(2) yields worse mobilities than BBBT(1). In ref. [139] interface trapping of charge carriers and high contact resistance are also suggested as factors limiting the mobility. Last but not least, it is worth emphasizing that the experimental publications only rarely state the directions in which charge transport was measured. In our simulations the mobility easily varies by a few orders of magnitude with changes in the transport direction. Especially for the two BBBT compounds this is a critical point, which is only briefly mentioned in the experimental publication [139].

7.4.3. Conclusions

A comparison of the alignment of nearest neighbors along the main transport directions and of the connectivity graphs for rubrene, indolocarbazole and the two BBBT derivatives is shown in fig. 7.15. One reason rubrene's mobility is higher than that of most other molecular crystals is certainly its almost defect-free two-dimensional percolation network. It not only has a clearly preferred transport direction with an extremely high coupling, but also a good secondary direction allowing the charge carrier to circumvent possible defects. BBBT(1) is very similar to rubrene in this respect with slightly worse coupling along its main and secondary directions, but therefore allowing three-dimensional transport. The fact it yields a lower mobility overall indicates that the additional percolation paths gained going from 2D to 3D transporting compounds do not influence the mobility strongly, because the existence of pathways to circumvent neighbors with low transfer integrals is more important than their number. BBBT(2) in contrast to the other crystals transports only in one direction. The influence of neighbors with low coupling therefore is substantially higher and is responsible for a factor of three difference in mobility depending on the starting position of a charge in a snapshot. This is because a charge carrier is confined a single column throughout and the corresponding mobility is limited by the lowest rate present leading to significant differences between columns. Indolocarbazole, despite its extremely low transfer integrals between neighbors, gives higher mobilities than one might expect due to isotropic coupling strengths and very narrow transfer integral distributions.

We would like to stress the fact that especially in the main coupling directions the approximations made in the hopping approach no longer hold and diffusion limited by thermal disorder should be used. The transfer integral distributions are independent of the theoretical picture, however, and statements about dimensionality and connectivity are fully valid. The actual KMC simulations are done to reveal the relative effect of disorder on mobility without the expectation that absolute mobility values will match experimental ones. However, regarding the low coupling directions hopping transport is the appropriate picture to use. Thus in a treatment combining wave-like and hopping transport we would expect mobilities along the weak transport directions to remain almost unchanged but to increase along the strong coupling directions.

In conclusion, the general concept that the best coupling and thus the best transport properties will be found for closely packed, cofacially aligned molecules should be reconsidered. Certainly, the transfer integral is at its absolute maximum for two molecules in cofacial align-

7. Organic Crystals

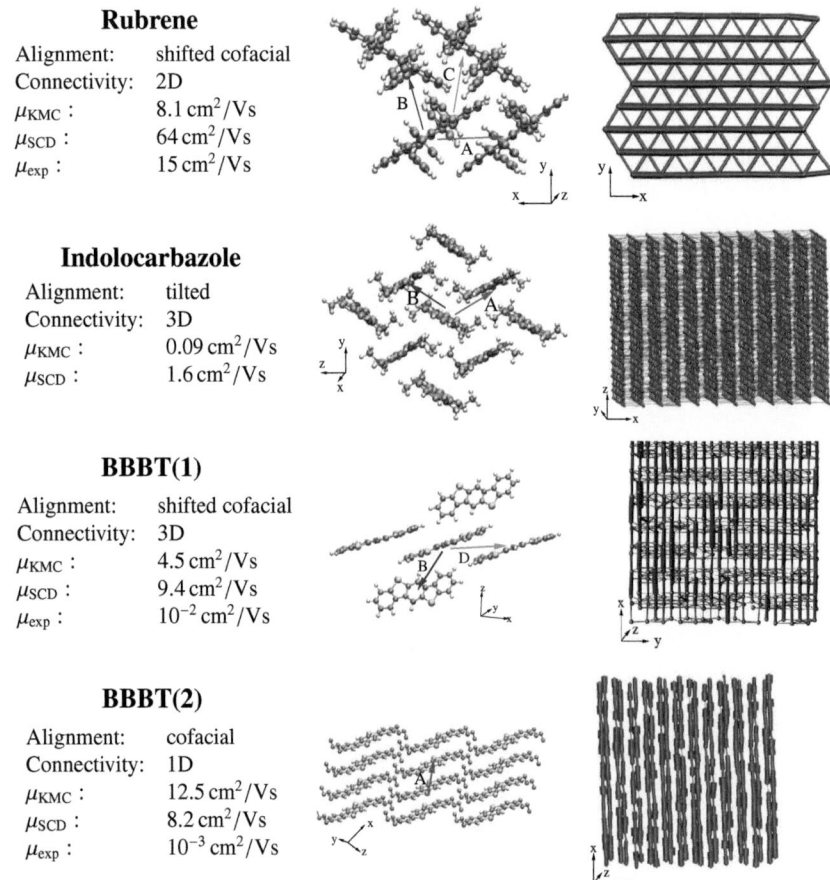

Fig. 7.15.: Comparison of the nearest neighbor alignment along the main transport direction and the connectivity in the system for rubrene, indolocarbazole and the two BBBT derivatives. Note that the grey spheres in the connectivity graphs, representing the centers of mass, are of the same size for all systems while the size of the bonds between them is proportional to corresponding the transfer integral.

ment only 3.5 nm apart and therefore the mobility along this direction is also at its maximum. However, the resulting morphology is likely to allow only one-dimensional transport thereby becomes extremely prone to defects. Such one-dimensional crystal morphologies are unlikely to be defect-free and yield good mobilities. Instead shifted cofacial alignment, as it is seen for BBBT(1) and rubrene and is illustrated in fig. 7.16b, is the far better alternative. While the transfer integrals will be lower due to smaller spatial overlap between molecules, the distance d travelled by a charge upon moving from one molecule to the other is increased thereby compensating for most of the loss in coupling. Most importantly, the shifted alignment allows two-dimensional transport. The resulting percolation network significantly reduces the influence of disorder by enabling charge transport despite a few defects.

Fig. 7.16.: Comparison of (a) cofacial alignment as in BBBT(2) and shifted cofacial alignment as in BBBT(1) and rubrene. While coupling is better in case (a), the charge travels a larger distance $d_b > d_a$ upon hopping in case (b) compensating for most of the loss in coupling. The main advantage of (b) over (a) is that it allows 2D and not just 1D transport and is thus less effected by disorder and defects.

7.5. Charge transport via semi-classical dynamics

The high temperature limit of Marcus theory is not applicable for strong couplings and good transport, i.e. in case of rubrene or for the main transport direction in the BBBT crystals, since the transfer integral J is no longer significantly smaller than the reorganization energy Λ (see sec. 4.4.2). In this situation an alternative approach is more appropriate, namely diffusion limited by thermal disorder and thus semi-classical dynamics (see sec. 4.3.1). The one-dimensional array of molecules is chosen such that it corresponds to the main transport direction within the crystal, e.g. direction A for rubrene. The separation between the sites d equals the average distance between nearest neighbors of that type. The average transfer integral between sites is the configurational (ensemble) average along that direction, the standard deviation is equal to the width of the $J(t)$ distribution for neighboring pairs with a similar average. This ensures that we take into account only dynamic and not static disorder. The discrete cosine transform of the autocorrelation of $J(t)$ is averaged over five to seven different pairs to yield the characteristic slow vibrational frequency $\omega^{(2)}$ in the system. The cosine transforms for all transport directions in indolocarbazole as well as for the main direction of both BBBT compounds are shown in fig. 7.17. For Rubrene refer to [75]. The fast vibrational frequency $\omega^{(1)}$ is chosen to be that of C-C bond fluctuations in phenyl rings and is thus the same for all systems (see again ref. [75]). The Peierls and Holstein coupling constants were calculated as described in sec. 4.3.1 using the reorganization energies computed based on single molecules optimized in vacuum using B3LYP/6-311G(d.p). The resulting input parameters used in the SCD simulations for all systems are summarized in tab. 7.2. All simulations were run at 300 K, the integration time step was chosen to be 0.0125 fs and each simulation consists of 600,000 steps. The wave function coefficients are written to file every 1000 steps. The simulations are repeated for 100 starting wave functions and the resulting random mean square displacement is Boltzmann averaged using the energy eigenvalue of the initial wave function for the weighting. Resulting mobilities are shown in the last row of tab. 7.2.

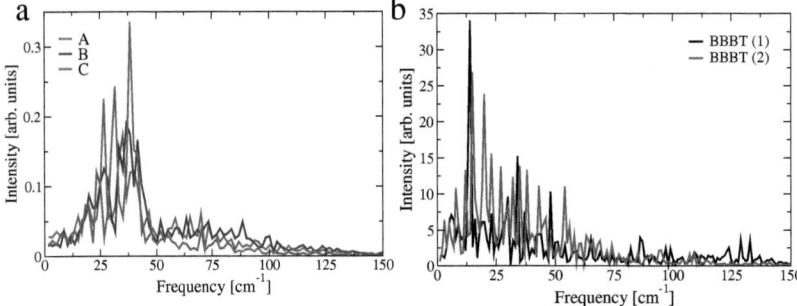

Fig. 7.17.: Discrete cosine transform of the autocorrelation function of $J(t)$ for (a) all directions in Indolocarbazole averaged over five neighboring pairs each and for (b) the main direction of BBBT(1) and (2) averaged over seven neighboring pairs.

As in experiments, rubrene shows the highest mobility followed by BBBT(1) and BBBT(2) with indolocarbazole having the lowest by far. Other than the high coupling between neighboring sites, rubrene also has the advantage that the molecules are rather large and due to the shifted cofacial alignment the distance between neighbors along the main coupling direction is almost twice that of BBBT(2). So despite the fact that average coupling is better in BBBT(2), rubrene shows the higher mobility, because the charge travels further between molecules. That BBBT(1) and (2) have roughly the same SCD mobilities may also be explained by this effect. The values of the mobility are also substantially higher than those predicted by Marcus rates, which is to be expected, since hopping transport is the slow limit of small polaron transport. The large relative differences between rubrene and BBBT are due to the different vibrational frequencies, since center of mass distances and reorganization energies are also taken into account in KMC. In Rubrene, molecular center of mass vibrations, described by the frequency $\omega^{(2)}$, are more than twice as fast as in BBBT, thus the lifetimes of small coupling values are reduced accordingly. Because all systems show almost no static disorder in the MD simulations, we expect the SCD results to hold more weight here than in the case of BTBT. Nevertheless, especially in the case of BBBT(2), it is clearly an overestimation of the mobility, since it is practically impossible to grow defect-free crystalline structures experimentally.

parameter	value			
$m^{(1)}$ [amu]	6			
$\omega^{(1)}$ [cm^{-1}]	1400			
	Indolocarbazole	Rubrene	BBBT(1)	BBBT(2)
$\lambda^{(1)}$ [cm^{-1}/Å]	25328	21200	18500	18500
J [cm^{-1}]	113	1090	312	−1126
$m^{(2)}$ [amu]	344	532	290	402
$\omega^{(2)}$ [cm^{-1}]	18	50	14	20
$\alpha^{(2)}$ [cm^{-1}/Å]	73	3980	417	2107
d [nm]	0.619	0.714	0.926	0.466
μ_{SCD} [cm^2/Vs]	1.57	64.3	9.43	8.15
μ_{KMC} [cm^2/Vs]	0.09	8.14	4.53	12.51
μ_{exp} [cm^2/Vs]	-	15	10^{-2}	10^{-3}

Tab. 7.2.: SCD input parameters used for the various crystals along with the resulting mobility μ at 300 K. For comparison, the mobilities along the main transport direction obtained by KMC simulations using Marcus rates as well as the experimental mobilities obtained by OFET measurements are also given.

7.6. Discussion

We have shown that indeed in case of ordered molecular crystals, except for those with very low couplings as indolocarbazole, the assumptions made in the derivation of the Marcus rates no longer hold, because the couplings along the main direction are of the same order as the reorganization energy. Low couplings only appear perpendicular to the main transport directions and are thus less influential on the charge transport properties. Nonetheless, agreement between rate-based description and small polaron transport are surprisingly good and insights could be gained on the influence of disorder in percolation networks of all dimensions.

Rubrene has been shown to be more favorable for charge transport than the other crystals in every respect: it has a low reorganization energy, a shifted cofacial alignment along the main direction of transport as well as reasonable coupling perpendicular to it for 2D transport and last but not least the highest center of mass vibrational frequency, limiting the lifetime of unfavorable alignments between neighbors.

The comparison of crystals with charge percolation networks of different dimensionality has shown that 1D transport is to be avoided due to a strong dependence on defects while it appears that 3D transport offers no significant improvement over 2D transport. Thus, since side chains improve the solubility and thereby make molecules more suitable for solution processing, soluble 2D transporting compounds should be favored over insoluble but 3D transporting compounds.

We have also shown that the often desired cofacial alignment between molecules yields the best coupling between nearest neighbors, but does not necessarily correspond to the highest mobility morphology, because cofacially aligned molecules tend to permit transport in only one dimension. We believe that shifted cofacial alignment is a more suitable arrangement to maximize the mobility, because it allows two dimensional transport and the reduction in coupling is compensated by the large distance a charge travels between molecules.

For molecules with a rigid fused-ring, linearly conjugated structure, shifted cofacial alignment only brings the benefits highlighted above, when the molecules are shifted along their linear direction. This is not possible, when the side chains are attached at these ends, since they will enforce a lamellar stacking of the molecules as in BBBT(2) and BTBT. Instead, attachment of the side chains perpendicular to the linearly conjugated core, as is the case for rubrene, ought to result in morphologies wit the highest mobilities. This is also supported by the high mobilities found for polymer field-effect transistors recently synthesized in such manner [145, 146].

8. Hexabenzocoronene

Among π-conjugated systems thermotropic discotic liquid crystals combine the fluidity of liquids, making them self-healing and solution-processable, with the orientational order of crystals, giving rise to anisotropic conductivity. The orientational order arises from the flat aromatic cores with aliphatic side chains, which self-assemble into columns. By adjusting the size and shape of the core as well as the side chains, compounds with different self-organizing and charge transport properties may be obtained. The effect of different alkyl side chains as well as liquid crystalline phases has already been studied in depth in our group [113, 92, 93, 94] for hexabenzocoronene (HBC), a hole-conducting discotic liquid crystal with a very high one-dimensional mobility of up to $1 \text{ cm}^2/\text{Vs}$ [147]. It is shown that the mobilities of different liquid crystalline phases measured by the pulse-radiolysis time-resolved microwave conductivity technique (PR-TRMC, see sec. 2.2.2) are well reproduced by simulations based on Marcus rates, i.e. rate-based simulations correctly describe charge transport on a local scale [93, 94]. The studies also reveal the dependence of mobility on the tail of the transfer integral distribution in 1D transporting systems, which is due to the fact that a charge carrier has to travel across every defect with no possibility to circumvent it.

To achieve high mobilities along columns of polyaromatic hydrocarbons, such as hexabenzocoronene, two distinct but related problems must be addressed: first, the compounds must self-organize into defect-free mesophases on a large scale. Second, the alignment of neighboring molecules in a column must be favorable for charge transport. Thus to allow rational design of such compounds, the best local molecular arrangement of the aromatic cores has to be found. The attached side chains then have to achieve this arrangement and in addition yield good self-organizing abilities. As shown in ref. [22] and fig. 8.1, for polyaromatic hydrocarbons the transfer integral between two neighboring cores has a maximum not only in cofacial configuration but also at a sixty degree twist. Side chains should hence be chosen in such a way that, besides offering processibility and self-assembling properties, they favor small intramolecular separations, since the transfer integral depends exponentially on the distance, and provide either a cofacial or a 60° twist molecular arrangement. One option are compounds with a six-fold symmetry, which will be in face-to-face arrangement both at 0° and 60°. However, steric repulsions between side chains are likely to increase the intermolecular separations and thereby widen the distribution of transfer integrals. The alternative are compounds with three-fold symmetry with a helical packing structure and 60° twist, which should allow for small intermolecular separations and thus ideal local arrangement for charge transport. That this is indeed the case has been shown for polyaromatic hydrocarbons with different symmetry [22] as well as for HBC derivatives with varying alkyl side chains [93, 94].

In this chapter we extend the studies of HBC derivatives by studying side chains, which are more complex than the linear or branched alkyl side chains treated previously. They are also chosen to have either six-fold or three-fold symmetry, such that they favor either cofacial alignment or a sixty degree twist between consecutive molecules in a column. The aim is to inves-

8. Hexabenzocoronene

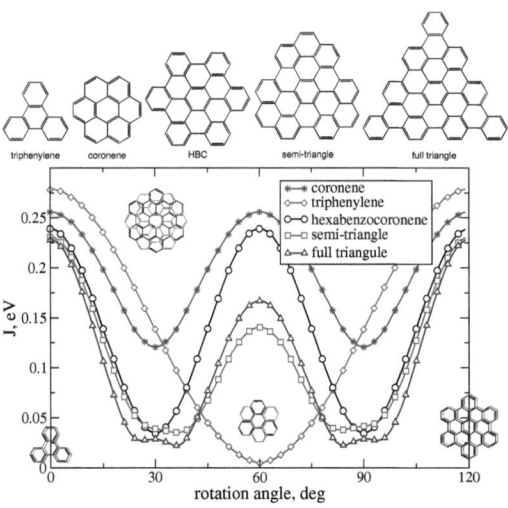

Fig. 8.1.: Absolute value of the transfer integral J as a function of the azimuthal rotation angle for several symmetric polyaromatic hydrocarbon cores. The separation was fixed to 0.36 nm. The insets illustrate face-to-face and staggered stacking of two typical disc-shaped molecules: triphenylene (staggered twisting angle is 60°) and HBC (staggered twisting angle is 30°). Taken from ref. [22].

tigate whether modifications of the chemical structure of the side chain, which influence the orientation of neighboring molecules, the columnar packing and the overall order in the system, can be utilized to significantly improve the charge transport properties. To this end four different side chains and their influence on morphology and hence mobility are studied. The predictions resulting from the MD simulations are intended to be compared with time of flight mobility measurements performed by F. Laquai and his group, however the latter are still work in progress.

8.1. Compounds and molecular dynamics simulations

The complete structures of the four analyzed compounds are shown in fig. 8.2. All side chains contain a phenyl ring, which for compounds (a) and (b) branches into three alkyl side chains attached to an oxygen atom or for compounds (c) and (d) branches into two plain alkyl side chains. In case of (a) and (c) the HBC core has six such side chains attached, in case of (b) and (d) only three. Due to the three-fold symmetry of the resulting molecules, they are designed to promote either cofacial alignment or a sixty degree twist between neighboring molecules during self-assembly.

The force field parameterization as well as the MD simulations were performed by Valentina Marcon. Force field parameters are given in refs. [113, 92, 93, 94]. In the starting configuration, all molecules were cofacially aligned. For equilibration purposes stochastic dynamic runs were performed for 2 ns with a time step of 2 fs constraining all bonds. Temperature coupling was done via Nosé Hoover thermostat, pressure coupling via Berendsen barostat. Production MD

8.1 Compounds and molecular dynamics simulations

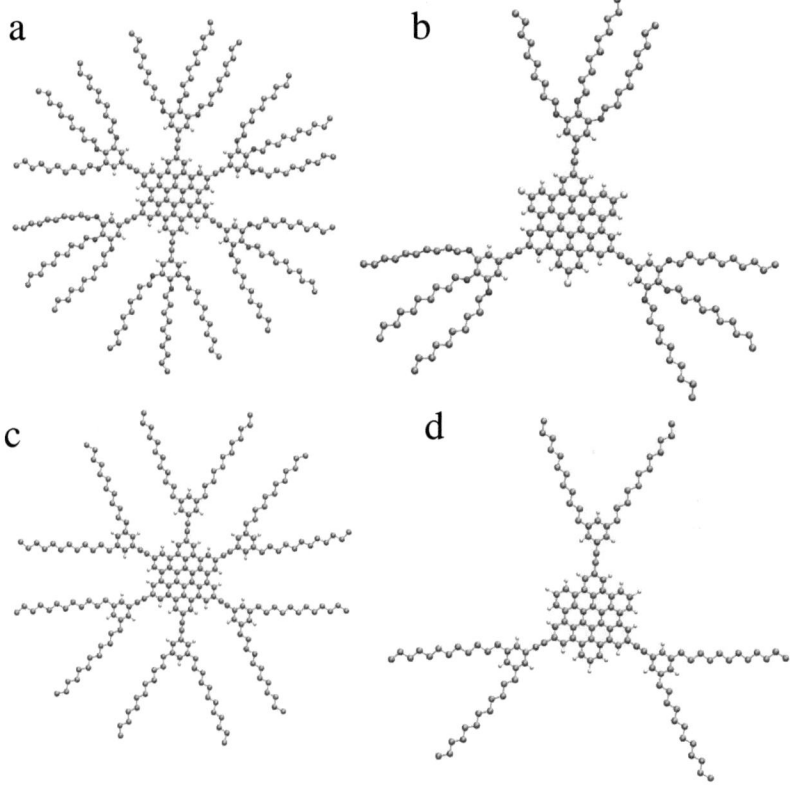

Fig. 8.2.: Structures of the four investigated HBC compounds. Note the three-fold symmetry for (b) and (d) and the six-fold symmetry for (a) and (c).

runs were performed in NPT ensemble for 10 ns. Bonds were constrained and temperature coupling was done via Nosé Hoover thermostat. The Parinello-Rahman barostat was used for pressure coupling. KMC simulations were performed based on snapshots taken in the last 0.4 ns of the production run. All simulations were done at 300 K.

8.2. Charge transport parameters

Since side chains do not contribute to charge transport, only the HBC core is used to compute the charge transport parameters. We focus solely on hole transport, because HBC is a hole-transporting organic semiconductor. The HOMO orbital of HBC is doubly degenerate. The two orbitals forming the HOMO are depicted in fig. 8.3a. They were calculated using the ZINDO method based on B3LYP/6-311G(d,p) optimized geometries. The internal reorganization energy for the cation was calculated using B3LYP/6-311G(d,p) for geometry optimization of a cation and a neutral molecule followed by an SCF calculation of the energies for the charged and neutral state. The result was $\Lambda = 0.099$ eV.

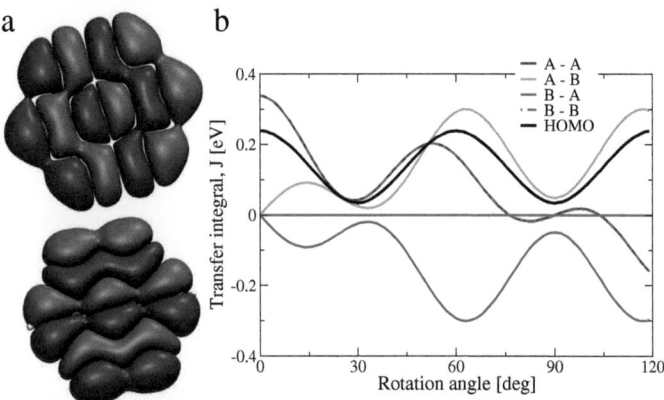

Fig. 8.3.: (a) The two orbitals A and B forming the degenerate HOMO of the HBC core. (b) Change in transfer integral upon rotation of two HBC cores 0.36 nm apart. The transfer integral is shown for all combinations of the orbitals A and B. Together they form the degenerate HOMO, whose transfer integral was calculated according to eqn. 8.1. Note that its transfer integral is always greater than zero.

The transfer integral is calculated using the molecular orbital overlap (MOO) method based on ZINDO orbitals [2]. The diabatic states are constructed using the frozen orbital approximation, where a cation has a hole occupying the HOMO orbital of a neutral molecule. Results of the computation for all transfer integral combinations of the two orbitals A and B forming the degenerate HOMO are shown in fig. 8.3b. For the KMC simulations the transfer integral J_{ij} between molecules i and j each with doubly degenerate orbitals A and B was calculated as

$$J_{ij}^2 = \frac{1}{4}\left(J_{A_i,A_j}^2 + J_{A_i,B_j}^2 + J_{B_i,A_j}^2 + J_{B_i,B_j}^2\right) \tag{8.1}$$

in accordance with [94, 96]. Eqn 8.1 takes into account all four transfer integral combinations between two different orbitals on two different molecules. The information on the sign of the transfer integral is lost in this procedure, but that is of no concern since only the square of the transfer integral enters in the calculation of the transfer rates. J_{ij} for two HBC cores rotated with respect to each other is also shown in fig. 8.3b. Due to the degeneracy J_{ij} is always larger than zero, which is not the case for any of the four constitutive transfer integrals.

8.3. Rate-based simulations of charge dynamics

Kinetic Monte Carlo simulations based on Marcus rates were performed for all compounds applying an external electric field of 10^7 V/m, which was aligned with the stacking direction (z) of the columns. As is illustrated in fig. 8.4a for compound (d), the large displacements of the molecules along the column do not show a clear nearest-neighbor distance. However, the distribution of transfer rates on logarithmic scale (see fig. 8.4b) reveals distinctive peaks corresponding to different orders of neighbors. Therefore, choice of a cut-off radius which takes into account second order neighbors has no effect on the simulation results, because any second-order neighbor transfer rate is orders of magnitude lower than that between nearest neighbors. In addition the side chains block all transport in the xy-plane, ensuring that there is only transport along the z-axis. In the end, the cut-off radius was chosen to be 1.2 nm throughout to prevent unphysical breaking of the column by not taking into account all nearest neighbors, i.e. the neighbors prior and after each molecule along the direction of the column (z).

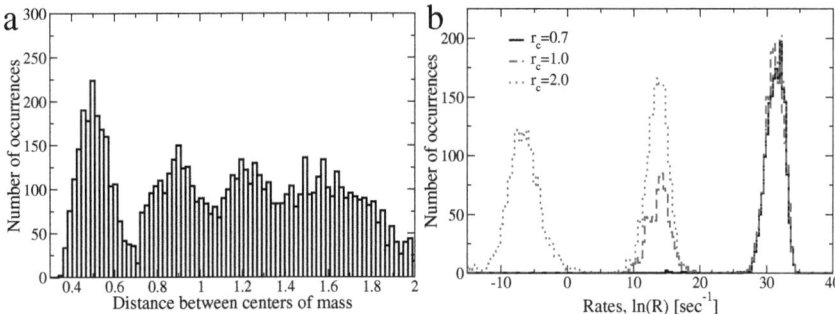

Fig. 8.4.: Distribution of (a) center of mass distances and (b) the logarithm of transfer rates used to choose an appropriate cut-off radius for compound (d). The distribution of transfer rates allows a clear distinction between first, second and higher order neighbors.

We performed charge transport simulations for three different cases: (i) calculating the transfer integral via eqn. 8.1 to take into account the degeneracy of the orbitals, (ii) additionally introducing electrostatic disorder and (iii) computing the mobility based on an ideal crystal structure without energetic disorder due to electrostatics. Electrostatic contributions were calculated by the procedure outlined in sec. 4.5.4. Upon equilibration all molecules remain in cofacial alignment along the column in case of (a) and (c), while they arrange into structures with a 60° twist between consecutive molecules in case of (b) and (d), i.e. every second molecule is identical

113

8. Hexabenzocoronene

μ_z	without electrostatic disorder	with electrostatic disorder	ideal structure
(a)	0.978	0.770	12.73
(b)	0.966	0.466	12.69
(c)	0.681	0.532	13.41
(d)	0.852	0.633	14.27

Tab. 8.1.: Mobilities in cm^2/Vs for the four different HBC compounds (a) to (d) with and without taking electrostatic disorder into account and for the ideal crystal. Taking into account electrostatic disorder yields the most realistic mobility value. The mobility based on the ideal structure shows the great decrease due to disorder.

along the column. Separation between molecules is ≈ 0.4 nm in all cases. The mobility in the ideal crystal is thus identical for all compounds and an order of magnitude higher than for the equilibrated systems, which is to be expected when introducing disorder in a one-dimensional conductor. To understand the difference in mobility for the four compounds, it is necessary to take a closer look at the corresponding equilibrated morphologies.

An impression of the difference in the equilibrated morphologies due to different side chains is given in fig. 8.5, where the ordering of the HBC cores in a single representative column is shown for each system. The molecules are tilted for (a) and the centers of mass deviate strongly from the initially well-aligned columns for (b) and (d). Deeper insights into the structure may be obtained by the radial distribution functions and nematic order parameters.

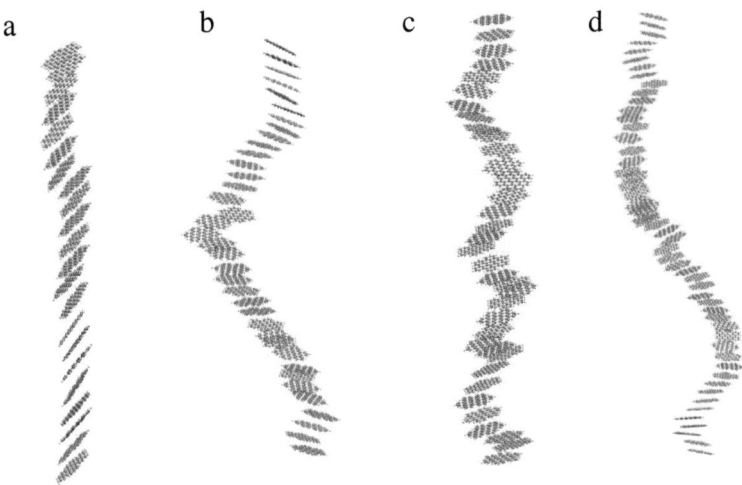

Fig. 8.5.: HBC core alignment in single columns taken from equilibrated MD simulations for systems (a) to (d).

Radial distribution functions

The radial distribution function (RDF) $g(r)$ gives the normalized probability of finding a molecule within a certain distance from another. In the z-direction it is computed as:

$$g_z(r) = \frac{2}{VN} \sum_{\text{columns}}^{N_c} \sum_{i=0}^{N_c} \sum_{j>i} f(\mathbf{r}, \mathbf{n} \cdot \mathbf{r}_{ij}) \qquad (8.2)$$

Here N is the total number of molecules, N_z the number of molecules per column, $\mathbf{r}_{ij}$ the distance between centers of mass of molecules i and j and f is a binning function, i.e. $f = 1$ if $\mathbf{n} \cdot \mathbf{r}_{ij} \in [r \pm \Delta/2]$, where Δ is the bin size (400 in this study), otherwise $f = 0$, the volume of the bin is given by $V = \Delta L_x L_y$ with L_x and L_y being the simulation box lengths. Averaging was first performed for every column and then averaged over all columns, since they can diffuse with respect to each other. The first maximum thus corresponds to nearest neighbors along the column and the existence and broadness of further maxima indicates the amount of disorder. No disorder corresponds to equally spaced delta functions placed at multiples of the nearest neighbor distance. In the xy-plane the RDF monitors the order in the hexagonal alignment of the columns. Both RDFs for all systems are presented in fig. 8.6. The systems (a) and (c) show a better order in the xy-plane with a pronounced maximum at ≈ 3.8 nm. Because they have six instead of just three side chains, this leaves little space for motions in direction of the side chains leading to a better arrangement of the columns with respect to each other. In the z-direction the (b) and (d) systems remain more closely stacked with a spacing of the maxima of roughly 0.4 nm. The cores with less side chains are able to approach each other more closely due to the 60° twist between consecutive molecules, reducing the interference of side chains. The long range order is best in (a) with more than ten well distinguishable maxima, while it is worst in (c), where features diminish significantly after the 2nd maximum.

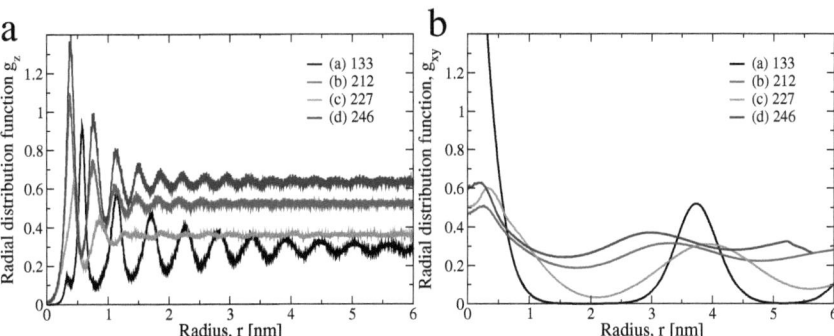

Fig. 8.6.: Radial distribution functions (a) along the column (z) and (b) perpendicular to it (xy).

Order parameters

To compute the orientational order parameter only the aromatic cores of the molecules are considered. Each core i is described by its normal vector $\mathbf{u}^{(i)}$. The orientational order tensor of

8. Hexabenzocoronene

the entire system may then be calculated as:

$$S_{\alpha\beta} = \left\langle \frac{1}{N} \sum_{i=1}^{N} \left(\frac{3}{2} u_\alpha^i u_\beta^i - \frac{1}{2} \delta_{\alpha\beta} \right) \right\rangle, \tag{8.3}$$

where N is the number of molecules in the system and $<\cdots>$ denotes the time average. Diagonalization of the tensor yields the order parameter S, which is the largest eigenvalue of $S_{\alpha\beta}$, and the director $\mathbf{n}$, which is the corresponding eigenvector and thus represents the average orientation of the molecules. $S = 1$ corresponds to a perfect molecular alignment, where all normal vectors $\mathbf{u}^{(i)}$ are parallel to each other, $S = 0$ implies isotropic angular distribution of the normal vectors. The order parameter and director for each system are listed in tab. 8.2. average distance between the cores along the column is too high to yield good electronic coupling. The

System	S	n_x	n_y	n_z
(a)	0.963	0.49	0.44	0.66
(b)	0.898	-0.06	0.02	0.98
(c)	0.948	-0.05	-0.04	0.95
(d)	0.892	0.03	-0.04	0.97

Tab. 8.2.: Order parameter S and components of the director $\mathbf{n}$ for all systems. The order parameter is close to one for the systems with six side chains, i.e. (a) and (c), and slightly worse for those with three. All directors are well-aligned with the z-axis (the stacking direction of the column) except for (a), where molecules are strongly tilted.

The ordering of the systems with oxygens in the side chains is superior to those without, i.e. (a) is always more ordered than (c) and (b) more ordered than (d). This might be due to dipoles induced by the negative partial charge on the oxygen and positive partial charges on the surrounding carbons, which cause stronger interactions between neighboring molecules. (a) has the best ordering overall leading to the highest mobility. It is followed by (b) and (d), where the cores approach each other more closely than in the rather disordered (c), which shows the lowest mobility. Electrostatic disorder reduces all mobilities, but not in an identical fashion. The mobility drops more strongly for the compounds with three side chains than for those with six. In ref. [94] it is shown that the energy difference between two neighboring HBC cores due to electrostatics is non-zero only if the molecules are tilted with respect to each other. Because the order parameter is higher for compounds (a) and (c), on average the tilt between neighboring molecules is less and they are thus less effected by electrostatics. Overall, compound (a) with six oxygen-containing side chains retains its superiority.

8.4. Conclusions

The modifications of the HBC side chains could be linked to changes in the corresponding morphologies and resulting mobilities. However, the neither side chain design present here yields significantly higher mobilities than the others and all mobilities are of the same order as for the alkyl side chains already presented in ref. [93, 94].
For the compounds (a) and (c) the molecular ordering along the columns due to the space-filling nature of their side chains is very good. However, the average distances between the conjugated

cores along the columns is too high to allow for strong electronic coupling. For compounds (b) and (d) the molecules align in a helical structure with a 60° degree twist, which allows a close approach between neighboring molecules. The drawback is that there is too much disorder due to molecular motions perpendicular to the columnar stacking direction, which leads to disorder in the transfer integral distributions and thus low transfer rates in the columns. The molecules also have enough space to tilt in irregular fashion with respect to each other, thereby increasing the effect of electrostatic disorder and reducing the mobility. However, the influence of electrostatic disorder in all systems is low and thus not an important parameter in columnar phases of HBC. Finally, the only structural modification of the side chains with solely positive effect on the ordering and mobility appears to be the addition of oxygen atoms.

We thus conclude that modification of the side chains is unlikely to improve charge mobility by orders of magnitude. Side chains chosen such that they provide either a cofacial or a 60° twist molecular arrangement yield similarly high mobilities no matter what the chemical details.

9. Discussion

Understanding the connection between chemical structure and charge transport properties in organic semiconductors is necessary to allow significant improvements of charge mobility and thus device performance in the future. While the general understanding of charge transport within band and disorder models has substantially improved in the past, the precise relations between chemical structure, bulk and interface morphology and device performance are up to this point mostly unknown.

Quantum chemical studies have helped to identify the orbitals relevant for charge transport and in combination with the hopping formalism and Marcus theory also explained the dependence of the transfer integral on molecular alignment. The latter shows the importance of the sign of the wave function and the resulting bonding-antibonding pattern between neighboring molecules. Theoretical physics and computer simulations have successfully described temperature, charge carrier density and electric field dependence of mobility in organic semiconductors. In this work we additionally used molecular dynamics simulations to capture the realistic molecular motion and thereby gain access to static and dynamic disorder by analysis of corresponding transfer integral distributions and resulting percolation pathways. Previous studies performed in our group have been able to uncover the dependence of one-dimensional transport in hexabenzocoronene on the tail of the transfer integral distributions. They have also allowed improvement of charge transport by choice of core and side chain morphologies aligning molecules at a 60° twist instead of attempting to achieve cofacial alignment. In this work we have both continued the study of columnar systems with carbazole macrocycles and hexabenzocoronene derivatives as well as performed a comparative study of semi-classical dynamics and rate-based simulations for organic crystals such as BTBT, rubrene, indolocarbazole and BBBT. For the understanding of charge transport in organic crystals we have developed new ways of visualizing transfer integral distributions and their effect on charge percolation. Not only were we able to connect transfer integral distributions with their corresponding neighboring pairs in the crystal, but we have also proposed the use of connectivity graphs for the visualization of charge percolation, and cluster analysis, yielding a reasonable approximation for an effective transfer integral value and also offering a visualization for the dependence of cluster size on the strength of the coupling between neighbors. While neither rate equations nor SCD are capable of predicting absolute values of mobilities, we were certainly able to gain a qualitative understanding of charge transport and to give precise advice on how to modify chemical structures of molecules and material morphologies to yield higher mobilities.

9.1. Charge transport in columnar systems

The dominance of the lowest transfer integrals over the transport along one-dimensional columns is a severe limitation for columnar systems. This limitation is overcome for carbazole macrocycles. Since their side chains are facing inwards, molecules of neighboring columns are able to approach each other closely, thus allowing for transport in three dimensions. In the traditional molecular design of discotics, side chains face outwards thus blocking all possibility of other than one-dimensional transport. The three-dimensionality enables charge carriers to circumvent defects along columns and improves the mobility by orders of magnitude.

For hexabenzocoronene, chemically modified side chains with six-fold and three-fold symmetry were compared to see whether the chemical details of the side chains significantly influence charge transport in discotic liquid crystals. This does not appear to be the case. As long as either cofacial alignment or a twist of 60° between neighboring molecules along the column exists in the self-assembled morphology, charge transport is equally good no matter the chemical details. Additionally, a better ordering in the self-assembled morphologies is observed for side chains containing oxygen compared to those without.

9.2. Static disorder

To reduce the production cost of electronic devices, solution processing must replace high vacuum vapor deposition in the fabrication of crystalline organic layers. This requires soluble molecules, i.e. alkyl or similar side chains, and inevitably leads to more disorder in the final device morphologies. As an example of such a solution-processable crystal we analyzed BTBT. The slow dynamics of the side chains lead to static disorder on the timescale of charge transport. This results in broad transfer integral distributions. That this is not an effect of dynamic disorder could be proven by comparing the configurational with the time-dependent transfer integral distributions between specific neighboring pairs in the system. Even though the transfer integrals distributions in BTBT are centered at low couplings, the crystal retains a charge percolation network with sufficient connectivity not to hinder transport. Nonetheless the coupling of 0.022 ev found between all nearest neighbor pairs in the ideal crystal is reduced to an effective coupling of roughly 0.015 eV in the equilibrated case. For higher coupling the connectivity of the network breaks down according to cluster analysis. This also shows that theoretical approaches based on ideal crystal morphologies and effective coupling strengths can both result in misleading conclusions.

9.3. Dimensionality

The effect of going from one-dimensional to multi-dimensional transport is illustrated in the comparison between fully conjugated carbazole macrocycles, where only transport along the column is taken into account, versus three-dimensional intercolumnar transport, which improves the mobility by orders of magnitude. BTBT also illustrates how disorder in a two-dimensional system does prevent charge transport as in the one-dimensional case but instead only modifies the underlying connectivity network. In the chapter on organic crystals, we deal with one-, two- and three-dimensional charge percolation networks and show that even very high average coupling along one direction is not capable to compensate for the loss of dimensionality (BBBT with C_4H_9 side chains). It is thus preferable to have two good coupling directions rather than to have a single excellent one with no alternative directions to circumvent defects. Regarding the difference between two- and three-dimensional networks, it appears that no major improvements are to be expected by going from 2D to 3D percolation networks. It is important for a system to have alternative pathways to circumvent defects, not how many such pathways there are. Thus it is acceptable to have one non-conductive direction, for example that of the side chains required for solubility.

9.4. Ideal morphologies

The highest electronic coupling between two molecules is found in cofacial alignment. Hence it is often believed that the highest mobilities will also be achieved in morphologies where the molecules are cofacially aligned. However, cofacial coupling between molecules in one direction tends to reduce the electronic coupling in all other directions, thus creating an effectively 1D transporting system prone to defects. The highest mobilities today are observed in organic crystals such as rubrene, which features shifted cofacial alignment between neighbors along its main transport direction. While the coupling is reduced compared to perfect cofacial alignment, it is still strong and the fact that the charge travels a greater distance upon hopping compensates for part of the loss. While increasing the distance between cofacially aligned molecules leads to an exponential drop in coupling, the effect of shifting the molecules with respect to each other is more complex and depends on the bonding-antibonding pattern of the orbitals and thus on the underlying chemical structure. Most importantly, shifted cofacial alignment gives rise to two-dimensional transport, allowing charges to circumvent weakly coupled neighbors and thus decreasing the influence of defects. In case of linearly conjugated molecules such as BTBT, BBBT and rubrene, the molecules must be shifted along their linear axis with respect to each other to enable high electronic coupling in shifted cofacial alignment. Therefore, we propose to attach the side chains perpendicular to the linear axis of the core to allow such structures to self-assemble.

9.5. SCD vs. rate-based descriptions

Semi-classical dynamics is designed to treat materials with medium strength coupling, where neither fully delocalized nor fully localized descriptions are valid. It does so taking into account both diagonal and off-diagonal dynamic disorder. Its weakness is the one-dimensionality, which does not allow for a sensible treatment of static disorder and can only take into account one coupling direction in crystals. Nevertheless, the mobility and its temperature dependence obtained for rubrene are in agreement with experiments and show that diffusion limited by thermal disorder is a good description of charge transport in well-ordered organic crystals. Rate-based descriptions, on the other hand, have been shown to correctly describe hopping transport. Since in strongly coupled systems such as organic crystals the assumption of charge localization is invalid, rate equations predict too low mobilities. However, neighboring pairs with extremely weak coupling due to unfavorable alignment of neighboring molecules cannot be treated by any other approach. In addition the comparison of three-dimensional rate-based simulations for ideal and equilibrated systems gives information on the influence of static disorder. Thus in the case of crystalline systems dominated by static disorder, for example due to slow side chain dynamics such as BTBT, a rate-based description is advisable, since it is then important to capture the low rates correctly.

9.6. Future possibilities

A promising approach to treat charge transport in well-ordered organic crystals in the future is by two- or three-dimensional semi-classical dynamics taking its input parameters from MD simulations. It is the appropriate way to treat dynamic disorder and intermediate coupling strengths. Static disorder could be implemented by calculating the set of parameters for each neighboring pair on its own, i.e. analyzing the J(t) distribution for each pair separately instead of averaging over pairs of the same type as was done in this work.

It is also of great interest to improve the analysis tools for charge transport in given morphologies. Code is currently being written to display the residence time of electrons in certain locations, a visualization of the most commonly taken percolation pathways and local electric currents and finally a tool to find more complex traps, such as areas where the charge moves in a circle or constantly hops back and forth between two neighbors. It is also important to take into account the interaction between charges in simulations to allow multiple charges to travel at the same time and thereby take into account the charge carrier density.

To study the influence of defects in morphologies and not just static disorder, it is worthwhile to look at larger systems, using for example systematic coarse-graining, until defects appear. Charge transport could then be studied by reintroducing atomistic details into the final coarse-grained morphologies and using the methods described in this thesis.

A. Appendix

A.1. Quantum Chemistry

A.1.1. Variational principle

To find the ground state energy and wave function of a molecule, it is sufficient to look at the non-relativistic, time-independent Schrödinger equation

$$\mathcal{H}|\Phi\rangle = \epsilon|\Phi\rangle$$

where $\mathcal{H}$ is the Hamiltonian, Φ the multi-electron wave function and ϵ the energy of the system. It can be shown that for the hermitian operator $\mathcal{H}$ there exists an infinite set of exact solutions, labelled by α, to this eigenvalue problem. Namely:

$$\mathcal{H}|\Phi_\alpha\rangle = \epsilon_\alpha|\Phi_\alpha\rangle$$

Assuming the set of eigenvalues is discrete $\epsilon_0 \leq \epsilon_1 \leq \epsilon_2 \leq \cdots \leq \epsilon_\alpha \leq \cdots$, since $\mathcal{H}$ is hermetian, the eigenvalues are real and the corresponding eigenfunctions are orthonormal. Therefore the eigenfunctions also form a complete basis set and any function $\tilde{\Phi}$ satisfying the same boundary conditions may be expressed as a linear combination thereof

$$|\tilde{\Phi}\rangle = \sum_\alpha c_\alpha |\Phi_\alpha\rangle = \sum_\alpha |\Phi_\alpha\rangle \langle\Phi_\alpha|\tilde{\Phi}\rangle$$

These considerations suffice to prove the following theorem.

The Variational Principle. *Given a normalized wavefunction $|\tilde{\Phi}\rangle$ which satisfies the appropriate boundary conditions, the expectation value of the Hamiltonian is an upper bound to the ground state energy, i.e.: If $\langle\tilde{\Phi}|\tilde{\Phi}\rangle = 1$ then $\langle\tilde{\Phi}|\mathcal{H}|\tilde{\Phi}\rangle \geq \epsilon_0$.*

Proof. Begin by rewriting the normalization criterion in terms of the eigenfunctions of the Hamiltonian.

$$\langle\tilde{\Phi}|\tilde{\Phi}\rangle = 1 = \sum_{\alpha\beta} \langle\tilde{\Phi}|\Phi_\alpha\rangle \langle\Phi_\alpha|\Phi_\beta\rangle \langle\Phi_\beta|\tilde{\Phi}\rangle = \sum_{\alpha\beta} \langle\tilde{\Phi}|\Phi_\alpha\rangle \delta_{\alpha\beta} \langle\Phi_\beta|\tilde{\Phi}\rangle$$
$$= \sum_\alpha \langle\tilde{\Phi}|\Phi_\alpha\rangle \langle\Phi_\alpha|\tilde{\Phi}\rangle = \sum_\alpha |\langle\Phi_\alpha|\tilde{\Phi}\rangle|^2$$

Applying the same idea again to the other equation yields:

$$\langle\tilde{\Phi}|\mathcal{H}|\tilde{\Phi}\rangle = \sum_{\alpha\beta} \langle\tilde{\Phi}|\Phi_\alpha\rangle \langle\Phi_\alpha|\mathcal{H}|\Phi_\beta\rangle \langle\Phi_\beta|\tilde{\Phi}\rangle = \sum_\alpha \epsilon_\alpha |\langle\Phi_\alpha|\tilde{\Phi}\rangle|^2$$

A. Appendix

Since $\epsilon_\alpha \geq \epsilon_0$ for all α, this leads to:

$$\langle \tilde{\Phi}|\mathcal{H}|\tilde{\Phi}\rangle \geq \sum_\alpha \epsilon_0 \left|\langle \Phi_\alpha|\tilde{\Phi}\rangle\right|^2 = \epsilon_0 \sum_\alpha \left|\langle \Phi_\alpha|\tilde{\Phi}\rangle\right|^2 = \epsilon_0$$

□

So one way to find an approximation for the ground state energy and wave function is to start from a normalized trial function $|\tilde{\Phi}\rangle$ depending on a set of parameters and vary these parameters until the expectation value $\langle\tilde{\Phi}|\mathcal{H}|\tilde{\Phi}\rangle$ reaches its minimum. This minimum is then the estimate of the exact ground state energy.

A.1.2. Born-Oppenheimer approximation

To apply the theory outlined above it is necessary to know the Hamiltonian of the system of interest. For a molecule composed of M nuclei and N electrons with position vectors $\mathbf{R}_A$ and $\mathbf{r}_i$, respectively, define $r_{iA} := |\mathbf{r}_i - \mathbf{R}_A|$, $r_{ij} := |\mathbf{r}_i - \mathbf{r}_j|$ and $R_{AB} := |\mathbf{R}_A - \mathbf{R}_B|$. The Hamiltonian of the system, in atomic units, may then be written as follows:

$$\mathcal{H} = -\sum_{i=1}^{N} \frac{1}{2}\nabla_i^2 - \sum_{A=1}^{M} \frac{1}{2M_A}\nabla_A^2 - \sum_{i=1}^{N}\sum_{A=1}^{M} \frac{Z_A}{r_{iA}} + \sum_{i=1}^{N}\sum_{j>i}^{N} \frac{1}{r_{ij}} + \sum_{A=1}^{M}\sum_{B>A}^{M} \frac{Z_A Z_B}{R_{AB}}$$

Here, M_A is the ratio of the mass of nucleus A to the mass of an electron and Z_A is the corresponding atomic number. The Laplacians involve differentiations with respect to the coordinates of the ith electron and the Ath nucleus. The first term represents the kinetic energy of the electrons, the second that of the nuclei, the third term is the Coulomb attraction between electrons and nuclei and the last two terms are the repulsions between electrons and between nuclei, respectively.

Based on ideas published by Born and Oppenheimer in 1927, one can simplify this equation by assuming that, since nuclei are a lot heavier and therefore also move more slowly than electrons, the electrons are moving in a field of fixed nuclei. The first step of the approximation is to write the total wave function as product of a nuclear part $\Phi_{nuc}(\{\mathbf{R}_A\})$ and an electronic part $\Phi_{el}(\{\mathbf{r}_i\}; \{\mathbf{R}_A\})$:

$$\Phi(\{\mathbf{r}_i\}, \{\mathbf{R}_A\}) = \Phi_{el}(\{\mathbf{r}_i\}; \{\mathbf{R}_A\})\, \Phi_{nuc}(\{\mathbf{R}_A\}) \tag{A.1}$$

The semicolon indicates that the electronic part depends explicitly on the electronic coordinates $\mathbf{r}_i$ but only parametrically on the nuclear coordinates $\mathbf{R}_A$. In the second step, the cross terms arising from the nuclear kinetic operator acting on the electronic part of the wave function are ignored, i.e. $\sum_{A=1}^{M} \frac{1}{2M_A}\nabla_A^2 \Phi_{el}(\{\mathbf{r}_i\}; \{\mathbf{R}_A\})$ is taken to be zero. The electronic Hamiltonian may then be written as

$$\mathcal{H}_{el} = -\sum_{i=1}^{N} \frac{1}{2}\nabla_i^2 - \sum_{i=1}^{N}\sum_{A=1}^{M} \frac{Z_A}{r_{iA}} + \sum_{i=1}^{N}\sum_{j>i}^{N} \frac{1}{r_{ij}}.$$

To obtain the total energy for a given system of fixed nuclei, the constant nuclear repulsion must be added to the solution obtained from the electronic Hamiltonian $\epsilon_{tot} = \epsilon_{el} + \sum_{A=1}^{M}\sum_{B>A}^{M} \frac{Z_A Z_B}{R_{AB}}$.

A.1.3. Linear combination of atomic orbitals (LCAO)

As beautiful and simple as the variational principle may appear, it is hardly possible to find a function close to the lowest energy wave function by optimization of its functional parameters without starting from a good initial guess. The first idea that came to mind was to use the orbitals of hydrogen (1s ,2s ,2p...), which could be calculated analytically, for other atom types as well and construct molecular orbitals ϕ as a linear combination of atomic orbitals ψ (LCAO): $\phi = \sum_{i=1}^{N} a_i \psi_i$. The set of N functions ψ_i is the basis set and each function is associated with a coefficient a_i. Since the chosen atomic orbitals are not necessarily orthonormal, the energy of the guess wave function will be given by:

$$E = \frac{\int (\sum_i a_i \psi_i) \mathcal{H} (\sum_j a_j \psi_j) d\mathbf{r}}{\int (\sum_i a_i \psi_i)(\sum_j a_j \psi_j) d\mathbf{r}} \tag{A.2}$$

$$= \frac{\sum_{ij} a_i a_j \int \psi_i \mathcal{H} \psi_j d\mathbf{r}}{\sum_{ij} a_i a_j \int \psi_i \psi_j d\mathbf{r}} = \frac{\sum_{ij} a_i a_j \mathcal{H}_{ij}}{\sum_{ij} a_i a_j S_{ij}} \tag{A.3}$$

$\mathcal{H}_{ij}$ is called the resonance integral, S_{ij} the overlap integral. The latter is the spatial overlap of two basis functions, the former is less intuitive, but a combination of basis functions giving rise to large overlap integrals will also have large resonance integrals and $\mathcal{H}_{ii}$ corresponds to the energy of an electron occupying orbital i, i.e. its ionization potential. Minimization of the energy with respect to all involved coefficients leads to N equations for each of the N coefficients, namely:

$$\sum_{i=1}^{N} a_i(\mathcal{H}_{ki} - ES_{ki}) = 0 \quad \forall k \tag{A.4}$$

Such a set of equations only has a non-trivial solution when the determinant formed by the coefficients, here given by $\mathcal{H}_{ki} - ES_{ki}$, is zero. The corresponding equation type is called a secular equation. The result will be N different energy values E_j, with each energy corresponding to a different set of a_{ij}. Each set of coefficients will define an optimal wave function, the one corresponding to the lowest energy that of the ground state of the molecule.

A.1.4. Self-consistent field (SCF)

To make numerical computations feasible the Hartree Hamiltonian splits the Hamiltonian of the entire system into single electron Hamiltonians $\mathcal{H} = \sum_i h_i$, with h_i given by:

$$h_i = -\frac{1}{2}\nabla_i^2 - \sum_{k=1}^{M} \frac{Z_k}{r_{ik}} + V_i\{\mathbf{j}\} \tag{A.5}$$

The kinetic energy of each electron as well as the interaction with the nuclei are easily separable. However, the interaction with all the other electrons occupying the orbitals **j** has to be computed as follows:

$$V_i\{\mathbf{j}\} = \sum_{i \neq j} \int \frac{\rho_j}{r_{ij}} d\mathbf{r} \tag{A.6}$$

Here ρ_j is the probability (charge) density associated with electron j. Since the probability density is given by $\rho_j = |\psi_j|^2$ we require the desired wave functions as an input. To solve this difficulty, Hartree proposed the following self-consistent field (SCF) approach [148, 149, 150]: start with an initial guess for the wave function, evaluate the densities required for the one-electron Hamiltonians and solve the corresponding differential equations $h_i \psi_i = \epsilon_i \psi_i$. Use the resulting wave function to recalculate the electron densities and repeat to obtain a yet better set of ψ. Stop iteration when the difference between preceding steps falls below a somewhat arbitrary threshold criterion. Note that the Hamiltonian is still considered to describe non-interacting electrons, since each individual electron sees only a constant potential, which does not instantaneously react to the electron's motion.

A.1.5. Slater determinant

Finally, it is important to take into account the quantum mechanical nature of electrons. This requires the introduction of the spin quantum number into the formalism. The spin of the electron has only two eigenvalues $\pm\hbar/2$, its eigenfunctions are orthonormal and will be denoted by α and β here. In addition the Pauli exclusion principle, that no two electrons can be characterized by the same set of quantum numbers, must also be taken into account. The first consequence is that each orbital may only be occupied by two electrons of opposite spin. The second is that the electronic wave function must be antisymmetric, i.e. change sign whenever two electrons are interchanged. So the wave function for two electrons of like spin has to have the following form:

$$\Psi = \frac{1}{\sqrt{2}} [\psi_a(1)\alpha(1)\psi_b(2)\alpha(2) - \psi_a(2)\alpha(2)\psi_b(1)\alpha(1)] \tag{A.7}$$

Equation A.7 may be rewritten by use of the determinant as follows:

$$\Psi = \frac{1}{\sqrt{2}} \begin{vmatrix} \psi_a(1)\alpha(1) & \psi_b(1)\alpha(1) \\ \psi_a(2)\alpha(2) & \psi_b(2)\alpha(2) \end{vmatrix} \tag{A.8}$$

The convenient coincidence, that determinants possess exactly the antisymmetric properties desired for electronic wave functions, was first noticed and used by Slater [151] in the construction of the so-called Slater determinants (SD) as N-electron wave functions,

$$\Psi_{SD} = \frac{1}{\sqrt{N!}} \begin{vmatrix} \chi_1(1) & \chi_2(1) & \cdots & \chi_N(1) \\ \chi_1(2) & \chi_2(2) & \cdots & \chi_N(2) \\ \vdots & \vdots & \ddots & \vdots \\ \chi_1(N) & \chi_2(N) & \cdots & \chi_N(N) \end{vmatrix} \tag{A.9}$$

where χ is a spin-orbital, i.e. a product of the spatial orbital and the spin eigenfunction. Note that the rows of the determinant correspond to electrons, while the columns are the different orbitals. This description therefore corresponds to closed shell systems, i.e. systems where each orbital is either doubly occupied (spin up and spin down) or empty, with the lowest energy E_0 being the ground state and the highest energy E_N belonging to the highest occupied molecular orbital (HOMO). An even more compact notation than eqn. A.9 is $\Psi_{SD} = |\chi_1\chi_2\chi_3\cdots\chi_N\rangle$. Upon expansion of the Slater determinant every electron appears in every spin orbital. This is a manifestation of the indistinguishability of quantum particles. When evaluating the Coulomb repulsion between the electron clouds in two orbitals a and b, the classical Coulomb repulsion is reduced by K_{ab} for electrons of like spin. This is the so-called 'Fermi hole', reflecting the reduced probability of finding two electrons of like spin close to each other. For two electrons of the same spin it is given by $K_{ab} = \int \psi_a(1)\psi_b(1)\frac{1}{r_{12}}\psi_a(2)\psi_b(2)d\mathbf{r}_1 d\mathbf{r}_2$.

A.1.6. Restricted Hartree Fock

Fock was the first to propose the use of Slater determinants in the SCF procedure introduced by Hartree. It was Roothaan, however, who described the matrix equations necessary to use a basis set representation for molecular orbitals (MOs) and stated the underlying mathematical proofs [152]. It is this formalism, that we describe in the following and that is the basis of most quantum chemical methods in use today [153]. The energy of a closed shell system is given by

$$E = 2\sum_i H_i + \sum_{ij}(2J_{ij} - K_{ij}), \qquad (A.10)$$

where H represents the nuclear-field single electron energies, J_{ij} refers to the coulomb and K_{ij} to the exchange integrals between the molecular orbitals i and j. The precise mathematical definitions of these energies are:

$$H_i = \int \bar{\chi}_i \mathcal{H} \chi_i d\mathbf{r} \qquad (A.11)$$

$$J_{ij} = e^2 \int \frac{\bar{\chi}_i(\mu)\bar{\chi}_j(\nu)\chi_i(\mu)\chi_j(\nu)}{r_{\mu\nu}} d\mathbf{r}(\mu)d\mathbf{r}(\nu) \qquad (A.12)$$

$$K_{ij} = e^2 \int \frac{\bar{\chi}_i(\mu)\bar{\chi}_j(\nu)\chi_j(\mu)\chi_i(\nu)}{r_{\mu\nu}} d\mathbf{r}(\mu)d\mathbf{r}(\nu) \qquad (A.13)$$

Here ν and μ label different electrons. Note that $J_{ii} = K_{ii}$ and that due to the orthogonality of the two different spin functions K_{ij} is zero for electrons of opposite spin. In the quantum mechanical spirit we may define the corresponding operators J_i and K_i as follows:

$$J_i(\mu)\chi(\mu) = e^2 \left(\int \frac{\bar{\chi}_i(\nu)\chi_i(\nu)}{r_{\mu\nu}} d\mathbf{r}(\nu) \right) \chi(\mu) \qquad (A.14)$$

$$K_i(\mu)\chi(\mu) = e^2 \left(\int \frac{\bar{\chi}_i(\nu)\chi(\nu)}{r_{\mu\nu}} d\mathbf{r}(\nu) \right) \chi_i(\mu) \qquad (A.15)$$

$$\qquad (A.16)$$

A. Appendix

This is sufficient to define the Fock operator, which is the sum of the total electron interaction and all one-electron energies taking into account only the potential of the fixed nuclei, i.e.:

$$\mathcal{F} = \mathcal{H} + \sum_i (2J_i - K_i) \tag{A.17}$$

It can be shown by taking into account the condition that molecular orbitals have to remain perpendicular even when undergoing small variations, that for a closed shell system it is sufficient to solve

$$\mathcal{F}\phi_i = \epsilon_i \phi_i \tag{A.18}$$

The matrix form thereof used for numerical evaluation based on the LCAO approach (sec. A.1.3) is

$$\mathbf{Fc_i} = \epsilon_i \mathbf{Sc_i} \tag{A.19}$$

Where $\mathbf{c_i}$ is the vector of atomic orbital coefficients corresponding to the molecular orbital i. S is the matrix of overlap integrals between different MOs arising from the secular equation $\det(\mathbf{F} - \epsilon \mathbf{S})$ to find the minimum energy basis functions. The Fock matrix is given by:

$$F_{\mu\nu} = \left\langle \mu \left| -\frac{1}{2}\nabla^2 \right| \nu \right\rangle - \sum_k^{\text{nuclei}} Z_k \left\langle \mu \left| \frac{1}{r_k} \right| \nu \right\rangle + \sum_{\lambda\sigma} \mathbf{P}_{\lambda\sigma} \left[(\mu\nu|\lambda\sigma) - \frac{1}{2}(\mu\lambda|\nu\sigma) \right] \tag{A.20}$$

Here the lower-case Greek letters refer to the basis functions of which the MOs are built and the latter so-called "four-index integrals" correspond to the coulomb and exchange interaction (J_{ij} and K_{ij}). Between two electrons labelled one and two they are defined as:

$$(\mu\nu|\lambda\sigma) = \int\int \phi_\mu(1)\phi_\nu(1)\frac{1}{\mathbf{r}_{12}}\phi_\lambda(2)\phi_\sigma(2)d\mathbf{r}(1)d\mathbf{r}(2) \tag{A.21}$$

The density matrix $\mathbf{P}_{\lambda\mu}$ describes how strongly individual basis functions contribute to the many-electron wave function. Equation A.18 is the best known form of the Fock equations. They state that the molecular orbitals which correspond to the most accurate Slater determinants are all eigenfunctions of the Hermetian operator $\mathcal{F}$. Once the Fock equations have been solved numerically, it is worthwhile to consider the eigenvalue problem of the now known Fock matrix $\mathbf{Fc} = \epsilon \mathbf{Sc}$. The lowest n eigenvalues thereof correspond to the solutions of eqn. A.18 and hence the energies of the occupied ground-state orbitals of the molecule. Orbitals corresponding to higher energies are excited states thereof. All eigenvalues are termed Hartree-Fock orbital energies.

The actual evaluation is done based on a chosen basis set in a self-consistent manner. As explained in sec. A.1.4 the initial coefficients of the wave functions are guessed to allow calculation of the Fock matrix. The Fock equations are then solved and result in a new set of wave function coefficients. This procedure is repeated until convergence is achieved. Due to the one-electron nature of the Fock operator, any electron correlation other than exchange is ignored. Two approaches have been developed to go beyond Hartree-Fock. The first is semi-empirical with the idea in mind to simplify the Fock equations by fitting coefficients to experimental data and yet gain accuracy. The second is to continue the ab initio approach and treat the electron correlation.

A.1.7. Semi-empirical methods

There are two main interests in the development of semi-empirical methods. First, to reduce the computational time by approximating extremely expensive calculations and second, to closer approach the real solution of the Schrödinger equation by taking the electron correlation into account better. Computationally the most costly step of the Hartree Fock (HF) theory is the evaluation of the two-electron or four-index integrals appearing in the definition of the Fock matrix (eqn. A.20). The main problem is that for N basis functions there are N^4 such integrals to be evaluated. A priori estimation of the integral values would save a lot of time. Since the coulomb integrals measure the repulsion between electrons whose location is defined by basis functions, basis functions occupying regions of space far away from each other should have close to zero integrals. Therefore, a cut-off for the distance between electrons is introduced above which all coulomb integrals are taken to be zero. While this is a purely numerical approximation and a choice of infinity as cut-off recovers the exact result, the approximations leading to truly efficient methods require some chemical background and intuition. One such approximation done by most semi-empirical methods is to ignore all core electrons, since they are taken to be sufficiently invariant to different chemical surroundings. Remaining valence orbitals are described by Slater-type orbitals (STO), which follow from the analytical solution of the Schrödinger equation for hydrogen and have the following form:

$$\phi(r, \theta, \phi; \zeta, n, l, m) = \frac{(2\zeta)^{n+1/2}}{[(2n)!]^{1/2}} r^{n-1} e^{-\zeta r} Y_l^m(\theta, \phi) \tag{A.22}$$

where ζ is chosen based on the atomic number, n is the principal quantum number for the valence orbital and $Y_l^m(\theta, \phi)$ are the spherical harmonics depending upon the angular momentum quantum numbers l and m.

In the following, we shall focus on methods fully or partially neglecting molecular overlap computations. The method for replacing Fock or overlap matrix elements was described by Pople and co-workers in 1965 and is termed complete neglect of differential overlap (CNDO) [154, 155]. It introduces the following simplifications:

i. The basis set is formed from STOs, one per valence orbital.

ii. Overlap matrix elements are defined by $S_{\mu\nu} = \delta_{\mu\nu}$

iii. Two-electron integrals (eqn. A.21) are taken to be zero unless the two orbitals on each atom are identical, i.e.:

$$(\mu\nu|\lambda\sigma) = \delta_{\mu\nu}\delta_{\lambda\sigma} = (\mu\mu|\lambda\lambda) \tag{A.23}$$

A. Appendix

iv. The remaining integrals are taken to be

$$(\mu\mu|\lambda\lambda) = \gamma_{AB} \qquad (A.24)$$

where A and B are the atoms belonging to the basis functions and γ may either be computed from s-type STOs or treated as a parameters. An example for the latter case is the Pariser-Parr approximation [156], where $\gamma_{AA} = IP_A - EA_A$ with IA_A being the atomic ionization potential and EA_A the electron affinity.

v. One-electron integrals for diagonal matrix elements are defined by

$$\left\langle \mu \left| -\frac{1}{2}\nabla^2 - \sum_k^{\text{nuclei}} \frac{Z_k}{\mathbf{r}_k} \right| \mu \right\rangle = -IP_\mu - \sum_k^{\text{nuclei}} (Z_k - \delta_{Z_A Z_k})\gamma_{Ak} \qquad (A.25)$$

Here the basis function μ is centered on atom A. The energy associated with the diagonal matrix element is the ionization potential in case the number of valence electrons and the valence nuclear charge are equal, i.e. all atoms have a partial charge of zero. The delta functions simply correct for self-interaction of electrons.

vi. One-electron integrals for off-diagonal elements are given by

$$\left\langle \mu \left| -\frac{1}{2}\nabla^2 - \sum_k^{\text{nuclei}} \frac{Z_k}{\mathbf{r}_k} \right| \nu \right\rangle = \frac{(\beta_A + \beta_B)S_{\mu\nu}}{2} \qquad (A.26)$$

μ and ν are centered on atoms A and B, respectively. $S_{\mu\nu}$ is the overlap matrix and the β values are semi-empirical parameters originally fitted to reproduce experimental data.

Thanks to eqn. A.23 the computational cost is reduced from N^4 to only N^2 and the evaluation of the remaining integrals is extremely simplified. However, failure to distinguish between different types and orientations of orbitals lead to significant errors and also to the development of a new formalism named intermediate neglect of differential overlap (INDO) [157]. The main change was simply to take different values for unique one-center two-electron integrals. For an atom having only s and p orbitals this lead to five such integrals instead of just one as in CNDO: $(ss|ss) = G_{ss}$, $(ss|pp) = G_{sp}$, $(pp|pp) = G_{pp}$, $(pp|p'p') = G_{pp'}$ and $(sp|sp) = L_{sp}$. These parameters may either be regarded as free parameters or estimated from spectroscopic data. Ridley and Zerner were the first to carefully parameterize INDO specifically for spectroscopic problems [158]. The resulting parameterization has been coined ZINDO and is also the method we use to calculate the HOMO orbitals of optimized geometries and the overlap between them to save computational time.

A.1.8. Basis sets

Prior to the discussion of further advances of HF theory it should be noted that the choice of basis set used is of key importance, when intent on reducing computational time while still retaining chemical accuracy. In the development of basis function three main goals are to be achieved:

i. Minimization of the number of basis functions.

  ii. Choice of functional forms allowing efficient evaluation of all integrals appearing in the HF equations.

  iii. Chemical applicability, i.e. the basis functions should have a large amplitude in areas where the probability density of the electron is also high.

The obvious choice for basis functions are Slater orbitals, but since they defy analytical solution of the four-index integrals (eqn. A.21) the functional form was changed from e^{-r} to e^{-r^2}. This lead to the use of Gaussian-type orbitals (GTOs) with the following normalized, atom-centered Cartesian coordinate form:

$$\phi(x,y,z;\alpha,i,j,k) = \left(\frac{2\alpha}{\pi}\right)^{3/4} \left[\frac{(8\alpha)^{i+j+k} i! j! k!}{(2i)!(2j)!(2k)!}\right]^{1/2} x^i y^j z^k e^{-\alpha(x^2+y^2+z^2)} \tag{A.27}$$

α controls the width of the GTO and i, j and k are non-negative integers defining the shape of the orbital. When all three integers are zero, the GTO has spherical symmetry and corresponds to s-type orbitals. When only one of the indices is equal to one, the function has symmetry about a single Cartesian axis. Depending on the choice of the index this corresponds to p_x, p_y or p_z orbitals. If the sum of the indices is equal to two, they represent d-type orbitals and so on. The main problem arising from choosing GTOs over STOs is the change in shape. GTOs are smooth and differentiable at the nucleus instead of possesing a cusp and the reduction in amplitude with distance is also too rapid. To obtain better agreement between functional forms, many basis functions are defined as a linear combination of Gaussian functions to approximate the Slater shape. Such basis functions are referred to as contracted basis functions, while the individual Gaussian are named primitive Gaussians. A single basis function is defined by its contraction coefficients c and the exponents α of its primitives. The first to systematically determine optimal contraction coefficients for GTOs mimicking STOs were Hehre, Stewart and Pople [159]. The resulting basis set appears as STO-MG in todays basis set terminology, short for 'Slater Type Orbital approximated by M Gaussians'. STO-3G basis functions have been defined for most atoms in the periodic table. STO-3G is a single-ζ meaning there is one and only one basis function for each type of orbital from the orbital core up to valence. This is the absolute minimum size possible for a reasonable basis set. A way to increase the flexibility of a basis set is to decontract it. For example one might decontract STO-3G and describe each AO by two basis functions, one being the contraction of two primitives, the other being the remaining primitive. Such a basis set, with two functions per AO, is called a double-ζ basis set. Examples thereof are the "correlation-consistent polarized Core and Valence (Double/Triple/etc.) Zeta" sets cc-pCVDZ, cc-pCVTZ, etc. developed by Dunning and co-workers [160]. The fact that core orbitals are only weakly affected by bonding, while valence orbitals on the other hand may vary greatly, lead to the introduction of 'split-valence' or 'valence-multiple-ζ' basis sets. Here, core orbitals continue to be represented by single contracted basis functions, while valence orbitals are split into arbitrarily many. The most widely used split-valence sets were also developed by Pople et al. They include 3-21G, 6-21G, 4-31G, 6-31G and 6-311G. The first number refers to the number of primitives in the contracted core functions, the numbers after the hyphen indicate the number of primitives in the valence functions. Two numbers correspond to a valence-double-ζ, three to a valence-triple-ζ basis set.

Polarization

Molecular in contrast to atomic orbitals require more flexible mathematical functions. An illustrative example is ammonia (NH_3), which has a pyramidal minimum energy structure, but the one predicted by above basis sets would be planar. Adhering to the convention that basis functions are centered on atoms, the added basis functions correspond to one quantum number higher angular momentum than the valence orbitals. In the Pople basis functions such polarization functions are indicated by one or two stars. For the 6-31G basis, a single star, i.e. 6-31G*, refers to a set of d functions added to polarize the p functions. A second star implies p functions on hydrogen and helium. Today, explicit enumeration of the polarization functions used in parentheses is also commonly used. 6-31G* is then denoted as 6-31G(d) and 6-31G(3d2f,2pd) implies heavy atoms to be polarized by three sets of d and two sets of f functions as well as hydrogen atoms by two sets of p and one of d functions.

Diffuse functions

Molecular orbitals of loose supermolecular complexes, anions or excited states have a tendency of being more diffuse than standard MOs. To be able to treat such weakly bound electrons standard basis sets are often 'augmented' with diffuse basis functions. In the Pople basis sets diffuse functions are indicated by a plus in the basis set name. 6-31+G(d) states that one set of s and one of p functions with small exponents have been added. The diffuse functions usually have exponents a factor of four below the smallest valence exponent. In the Dunning basis set such functions are indicated by prefixing the names with 'aug'. For the calculation of electron affinities and acidities use of such functions is highly recommended.

A.1.9. Post Hartree Fock

The fundamental assumption governing Hartree Fock theory is that each electron moves in a static electric field created by all others. The main difference between the Hamiltonian and the Fock operator is therefore that the former returns the electronic energy of the many-electron system while the latter is only a set of all interdependent one-electron operators, which is then used to find the one-electron MOs and build the Slater determinant. It is impossible to improve on HF theory with a single determinant, so it immediately comes to mind to instead take a linear combination of multiple determinants, i.e.:

$$\Psi = c_0 \Psi_{HF} + c_1 \Psi_1 + c_2 \Psi_2 + \cdots \tag{A.28}$$

where the coefficients c_i reflect the weight of each determinant. The main error in HF stems from the correlated motion of each electron with every other. It is referred to as dynamical correlation. Empirically it was shown that in this case the HF wave function dominates the linear combination given by eqn. A.28. In case of near or exactly degenerate frontier orbitals (HOMO or LUMO), different determinants may have similar weights. To distinguish this case it was named non-dynamical correlation.

Multiconfigurational Self-Consistent Field Theory (MCSCF)

Assuming a doubly degenerate frontier orbital, only one of the two orbitals π_2 or π_3 can be filled in the calculation of the Slater determinant. Even when including the other orbital in the calculation, only the occupied orbitals are optimized in the SCF process since only they contribute to the energy. Hence there is no driving force to optimize virtual orbitals, they are solely made to be orthogonal to occupied MOs. For our two-dimensional example the obvious solution is to form the wave function as a linear combination of two wave functions, one with orbital π_2 and one with π_3 occupied:

$$\Psi_{MCSCF} = a_1 \left| \cdots \pi_1^2 \pi_2^2 \right\rangle + a_2 \left| \cdots \pi_1^2 \pi_3^2 \right\rangle \tag{A.29}$$

The coefficients a_1 and a_2 take care of the normalization and relative weighting. This approach is called Multiconfigurational Self-Consistent Field (MCSCF), because the orbitals are optimized for a combination of different configurations. Each term in the linear combination is termed configuration state function (CSF) and refers to the molecular spin state and the occupation numbers of the orbitals. The resulting variational algorithm has to find not only the ideal shape for each MO but also the correct weight for each CSF in the MCSCF wave function. The resulting orbitals do not have unique eigenvalues and thus the energy of an orbital cannot be discussed. How many different orbitals to include in these considerations depends on the chemistry. The resulting choice is the active space given by (m, n), where m is the number of electrons and n the number of orbitals. Distributing the electrons on the orbitals gives the possible number of CSFs:

$$N = \frac{n!(n+1)!}{\left(\frac{m}{2}\right)!\left(\frac{m}{2}+1\right)!\left(n-\frac{m}{2}\right)!\left(n-\frac{m}{2}+1\right)!} \tag{A.30}$$

Permitting all possible arrangements of electrons in the MCSCF expansion is referred to as choosing a complete active space (CAS), i.e. doing CASSCF. If one chooses to do CASSCF for all electrons including all orbitals in the complete active space the calculation is called full configuration interaction or full CI. A full CI with an infinite basis set would be the exact solution of the non-relativistic, Born-Oppenheimer, time-independent Schrödinger equation. Note, however, that already for methanol in 6-31G(d) basis the number of coefficients to optimize in full CI for this (14,38) system would be $2.4 \cdot 10^{13}$ despite the small basis set choice.

Configuration Interaction (CI)

To move from the enormous task of a full CI calculation to a feasible task one approach is to allow only a limited number of excitations. For this purpose we rewrite eqn. A.28 as

$$\Psi = c_0 \Psi_{HF} + \sum_{i}^{\text{occ.}} \sum_{r}^{\text{virt.}} a_i^r \Psi_i^r + \sum_{i<j}^{\text{occ.}} \sum_{r<s}^{\text{virt.}} a_{ij}^{rs} \Psi_{ij}^{rs} + \cdots \tag{A.31}$$

Here, i and j are occupied MOs in the HF wave function, r and s are virtual MOs in Ψ_{HF} and the other CSFs are generated by exciting an electron from the occupied (subscripts) to the virtual (superscripts) orbitals. Including single and double excitations leads to CISD, which scales as N^6, however.

A.1.10. Perturbation Theory

A completely different approach is to apply the well-known Rayleigh-Schrödinger perturbation theory to solve the Schrödinger equation. In general, perturbation theory deals with an operator $\mathcal{A}$, that may be rewritten as

$$\mathcal{A} = A^{(0)} + \lambda V \tag{A.32}$$

with $A^{(0)}$ being an operator, where the eigenfunctions are easy to find, V being the perturbing operator and λ the dimensionless parameter, which upon variation from zero to one maps $\mathcal{A}$ to $A^{(0)}$. If the perturbation is small the ground state eigenfunctions and eigenvalues may be expanded into a Taylor series in λ. Since Taylor series are well-known we will denote the expansions as follows: $\Psi_0 = \Psi_0^{(0)} + \lambda \Psi_0^{(1)} + \lambda^2 \Psi_0^{(2)} + \cdots$ and $a_0 = a_0^{(0)} + \lambda a_0^{(1)} + \lambda^2 a_0^{(2)} + \cdots$. $a_0^{(0)}$ is the eigenvalue for $\Psi_0^{(0)}$, the normalized ground-state eigenfunction for $A^{(0)}$ and the terms with superscripts (n) are the nth-order corrections to the zeroth order term. This allows us to write

$$(A^{(0)} + \lambda V)|\Psi_0\rangle = a|\Psi_0\rangle \tag{A.33}$$

Evaluating eqn. A.33 and sorting the resulting terms by powers of λ allows the calculation of the coefficients using $\langle \Psi_0 | \Psi_0^{(0)} \rangle = 1$ and $\langle \Psi_0^{(n)} | \Psi_0^{(0)} \rangle = \delta_{n0}$. The results for the first and second order corrections of the energy are

$$a_0^{(1)} = \langle \Psi_0^{(0)} | V | \Psi_0^{(0)} \rangle \tag{A.34}$$

$$a_0^{(2)} = \sum_{j>0} \frac{\left|\langle \Psi_j^{(0)} | V | \Psi_0^{(0)} \rangle\right|^2}{a_0^{(0)} - a_j^{(0)}} \tag{A.35}$$

The first to apply this theory to the complete Hamiltonian operator were Møller and Plesset in 1934. The resulting methods are termed MPn, where n is the order at which the expansion is truncated. We have used MP2 for the computation of dihedral parameters and the geometry optimization of single molecules. $H^{(0)}$ is chosen to be the sum of one-electron Fock operators

$$H^{(0)} = \sum_{i=1}^{n} f_i \tag{A.36}$$

where n is the number of basis functions. In addition $\Psi^{(0)}$ is the Slater determinant made up of the occupied orbitals. The eigenvalue of $H^{(0)}$ is then given by sum of the occupied orbital energies, since:

$$H^{(0)}\Psi^{(0)} = \sum_{i}^{\text{occ.}} \epsilon_i \Psi^{(0)} = a^{(0)}\Psi^{(0)} \tag{A.37}$$

The correction term V yielding the full Hamiltonian is:

$$V = \sum_{i}^{\text{occ.}} \sum_{j>i}^{\text{occ.}} \frac{1}{r_{ij}} - \sum_{i}^{\text{occ.}} \sum_{j}^{\text{occ.}} (J_{ij} - \frac{1}{2}K_{ij}) \tag{A.38}$$

The first term on the right hand side is the proper way to calculate electron repulsion, while the second term is how it is computed by summation over Fock operators for occupied orbitals. Note that the first order correction simply reproduces the Hartree Fock energy, i.e. $a^{(0)} + a^{(1)} = E_{HF}$. So the approach only makes sense starting from second order, the corresponding correction is

$$a^{(2)} = \sum_{i}^{\text{occ.}} \sum_{j>i}^{\text{occ.}} \sum_{a}^{\text{vir.}} \sum_{b>a}^{\text{vir.}} \frac{[(ij|ab) - (ia|jb)]^2}{\epsilon_i + \epsilon_j - \epsilon_a - \epsilon_b} \tag{A.39}$$

Here the terms in the nominator correspond to the four-index integrals defined by eqn. A.21. The sum of $a^{(0)}$, $a^{(1)}$ and $a^{(2)}$ is the MP2 energy. MP2 scales roughly as N^5, MP4 as N^7, where N is the number of basis functions. Note that perturbation theory is built on the assumption that the perturbation is small, but this is hardly the case for the electron-electron repulsion energy. Also in contrast to all other methods presented this far perturbation theory is not based on the variational principle. Therefore, the estimate for the correlation energy may be too large instead of too small.

A.2. Density Functional Theory (DFT)

A.2.1. Thomas Fermi theory

Since the early 1920s this theory was used to roughly compute the electronic energy based on electron density distributions. However, it did not predict chemical binding and therefore was of almost no use for questions of chemistry and material science. It did, however, consider interacting electrons moving in an external potential $v(\mathbf{r})$ and provided an implicit relation between $v(\mathbf{r})$ and the density distribution $n(\mathbf{r})$:

$$n(\mathbf{r}) = \gamma(\mu - v_{eff}(\mathbf{r}))^{3/2} \; ; \; \gamma = \frac{1}{3\pi^2}\left(\frac{2m}{\hbar^2}\right)^{3/2} \tag{A.40}$$

$$v_{eff}(\mathbf{r}) = v(\mathbf{r}) + \int \frac{n(\mathbf{r}')}{|\mathbf{r} - \mathbf{r}'|} d\mathbf{r}' \tag{A.41}$$

Here μ is the chemical potential and the first equation is based on the expression $n = \gamma(\mu - v)^{3/2}$ for the density of a uniform, non-interacting, degenerate electron gas in a constant external potential. The second term in the second equation is the classically computed electrostatic potential, also referred to as the Hartree potential v_H. Despite the fact that TF theory was a rough solution to the many-electron Schrödinger equation, it was unclear whether there was a strict connection between them and whether knowledge of the ground state density $n(\mathbf{r})$ alone uniquely determined the system. This mystery was solved by Hohenberg and Kohn [161].

A.2.2. The Hohenberg Kohn theorem

The Hohenberg Kohn theorem. *The ground state density $n(\mathbf{r})$ of a bound system of interacting electrons in some external potential $v(\mathbf{r})$ determines this potential uniquely.*

Proof. The proof is given for a non-degenerate ground state, but may easily be expanded to degenerate ground states as well. Let $n(\mathbf{r})$ be the non-degenerate ground state density of N electrons in the potential $v_1(\mathbf{r})$, corresponding to the ground state Φ_1 and the energy E_1. Then:

$$E_1 = \langle \Phi_1 | \mathcal{H}_1 \Phi_1 \rangle = \int v_1(\mathbf{r}) n(\mathbf{r}) d\mathbf{r} + \langle \Phi_1 | (T + U) \Phi_1 \rangle,$$

where $\mathcal{H}_1$ is the total Hamiltonian corresponding to v_1 and T and U are the kinetic and interaction energy operators. Note that the external potential v_1 is a multiplicative operator with respect to Φ. Now assume that there exists a second potential $v_2(\mathbf{r})$, *not* equal to $v_1(\mathbf{r})$ + constant, with ground state Φ_2, necessarily $\neq e^{i\theta}\Phi_1$, which gives rise to the same $n(\mathbf{r})$. Then

$$E_2 = \int v_2(\mathbf{r}) n(\mathbf{r}) d\mathbf{r} + \langle \Phi_2 | (T + U) \Phi_2 \rangle.$$

Since Φ_1 is assumed to be non-degenerate, the variational principle (sec. A.1.1) for Φ_1 gives rise to the following inequality:

$$E_1 < \langle \Phi_2 | \mathcal{H}_1 \Phi_2 \rangle = \int v_1(\mathbf{r}) n(\mathbf{r}) d\mathbf{r} + \langle \Phi_2 | (T + U) \Phi_2 \rangle \quad \text{(A.42)}$$
$$= E_2 + \int [v_1(\mathbf{r}) - v_2(\mathbf{r})] n(\mathbf{r}) d\mathbf{r}$$

Analogously

$$E_2 \leq \langle \Phi_1 | \mathcal{H}_2 \Phi_1 \rangle = E_1 + \int [v_2(\mathbf{r}) - v_1(\mathbf{r})] n(\mathbf{r}) d\mathbf{r} \quad \text{(A.43)}$$

where $\leq$ is used since non-degeneracy of Φ_2 was not assumed. Adding equations (A.42) and (A.43) leads to the contradiction:

$$E_1 + E_2 < E_1 + E_2$$

This shows that the assumption of the existence of a second potential $v_2(\mathbf{r})$, which is unequal to $v_1(\mathbf{r})$ + const but yet yields the the same $n(\mathbf{r})$, must be wrong. □

Since $n(\mathbf{r})$ determines both N and $v(\mathbf{r})$ it gives us the full $\mathcal{H}$ and N for the electronic system. Therefore $n(\mathbf{r})$ implicitly determines *all* properties of the system derivable from solving the Schrödinger equation for $\mathcal{H}$.

N- and V-representability

A question that may arise immediately linked to the HK theorem is how one knows, after having minimized a given density, whether this density is a density arising from an antisymmetric N-body wave function (N-representability) and whether it actually corresponds to the ground-state density of a potential $v(\mathbf{r})$ (v-representability). It has been shown that any non-negative function can be written in terms of some $\Phi(\mathbf{r}_1, \mathbf{r}_2, \cdots, \mathbf{r}_N)$, so N representability is not a problem. However, for v-representability no general solution exists, but the constrained search algorithm [162, 163] of Levy and Lieb shows that this is irrelevant for the proof above.

A.2.3. Applications of DFT

- Since $v(\mathbf{r})$ depends on a set of parameters - lattice constants or nuclei positions - the energy may be minimized with respect to these quantities yielding molecular geometries and sizes, lattice constants, charge distributions etc.

- Compressibilities, phonon spectra, bulk moduli (solids) and vibrational frequencies (molecules) may be obtained from the change in energy with respect to the corresponding parameters.

- Dissociation energies are calculated as the energy difference between a composite system (e.g. a molecule) and its constituent parts (e.g. individual atoms).

- Electron affinities and ionization energies correspond to the energy difference between the neutral ground state energy and the total energy of the system with one additional or one less electron. Note that within the local density and generalized gradient approximations the $(N+1)^{\text{st}}$ electron is too weakly bound or even unbound. The asymptotic potential obtained decays exponentially and not as $1/r$, so the binding of negative ions is strongly suppressed. Self-interaction corrections or other fully non-local functionals are needed to improve this behavior.

- Forces are calculated from the derivative of the total energy with respect to the nuclear coordinates using the Hellmann-Feynman theorem.

A.2.4. The Kohn-Sham equations

In principle it is sufficient to express the total energy of a system in terms of its density and minimize this functional to obtain the ground state energy, density and wave function. The total energy of a typical molecule or solid is given by:

$$E(n) = T(n) + U(n) + V(n) = T(n) + U(n) + \int n(\mathbf{r})v(\mathbf{r})\,\mathrm{d}^3r. \qquad (A.44)$$

A. Appendix

Here T is the kinetic energy of the electrons, U the potential energy due to the interaction between them and V the potential energy from an external potential v, usually the electrostatic potential of the nuclei, which are taken to be fixed in space (Born-Oppenheimer approximation, sec. A.1.2). This is the only potential that is easy to treat since it is a multiplicative operator. For the others a universal density functional is unknown at this time, but it is possible to treat them with the Kohn-Sham equations.

The idea of the Kohn-Sham approach is to reintroduce a special type of wave functions (single particle orbitals) into the formalism, to treat the kinetic and interaction energy terms. In this approach the energy is taken to be composed of the following terms:

$$E(n) = T(n) + U(n) + V(n) = T_s\{\phi_i(n)\} + U_H(n) + E_{xc}(n) + V(n) \tag{A.45}$$

Here the kinetic energy is split into two contributions:

$$T(n) = T_s(n) + T_c(n)$$

T_s stands for the kinetic energy of non-interacting particles of density n with s denoting 'single particle' and T_c being the remainder with c denoting 'correlation'. The interaction energy U is approximated by the classical electrostatic interaction or Hartree energy U_H. The new energy term E_{xc} stands for a correction due to exchange (x) and correlation (c) effects, i.e. is given by $E_{xc} = (T - T_s) + (U - U_H) = T_c + (U - U_H)$. It is often decomposed as $E_{xc} = E_x + E_c$, where E_x is the exchange energy due to the Pauli principle (antisymmetry) and E_c is due to correlations (T_c is then a part of E_c).

The kinetic energy of non-interacting particles T_c

For non-interacting particles, the kinetic energy is nothing other than the sum of the kinetic energies of each particle. Written in terms of wave functions that is:

$$T_s(n) = -\frac{\hbar^2}{2m} \sum_i^N \int \phi_i^*(\mathbf{r}) \nabla^2 \phi_i(\mathbf{r}) d^3 r \tag{A.46}$$

$T_s(n) = T_s[\{\phi_i(n)\}]$ is an explicit orbital functional, but only an implicit density functional, due to the dependence of the wave functions on the density. Because T_s is defined as the expectation value of the kinetic energy operator $\hat{T}$ with the Slater determinant arising from the density n, i.e. $T_s(n) = \langle \Phi(n) | \hat{T} | \Phi(n) \rangle$, all consequences of antisymmetrization (exchange) are described by employing a determinal wave function in defining T_s. Hence, T_c, the difference between T_s and T is a pure correlation effect.

The Hartree energy U_H

The classical electrostatic interaction energy (Hartree energy) as well as the mean field result U_H in terms of density is given by

$$U_H = \frac{q^2}{2} \int d^3 r \int d^3 r' \frac{n(\mathbf{r})n(\mathbf{r}')}{|\mathbf{r} - \mathbf{r}'|} \tag{A.47}$$

The exchange energy E_x

Since the exchange energy correction stems purely from the interaction potential, it may be written in terms of single-particle orbitals as:

$$E_x[\{\phi_i(n)\}] = -\frac{q^2}{2} \sum_{jk} \int d^3r \int d^3r' \, \frac{\phi_j^*(\mathbf{r})\phi_k^*(\mathbf{r}')\phi_j(\mathbf{r}')\phi_k(\mathbf{r})}{|\mathbf{r} - \mathbf{r}'|} \quad (A.48)$$

where a single term in the sum corresponds to the energy of exchanging molecule j located at $\mathbf{r}$ with molecule k located at $\mathbf{r}'$. No exact expression in terms of the density is known. The exchange energy therefore describes the energy lowering due to antisymmetrization, i.e. the tendency of electrons with like spin to avoid each other, giving rise to the so-called 'exchange-hole'. It also corrects the Hartree term (A.47) for the self-interaction (see eqn. A.49).

The correlation energy E_c

The correlation energy is the difference between the full ground-state energy obtained from the correct many-body wave function and the one obtained from the Hartree-Fock or Kohn-Sham Slater determinant. Recalling the interpretation of the wave function as probability amplitude, the product form of the many-body wave function corresponds to treating the probability amplitude of a many-electron system as a product of the probability amplitudes of individual electrons (the orbitals). This is only the same when the individual electrons are independent. Clearly this is not the case, so such wave functions neglect the fact that electrons try to avoid each other due to Coulombic repulsion. The correlation energy is therefore the additional energy lowering obtained in a real system due to the mutual avoidance of the interacting electrons. To understand the correlation for interaction energy, let us take another look at the corresponding operator written in two perfectly equivalent ways:

$$\hat{U} = \sum_{i<j} \frac{q^2}{|\mathbf{r} - \mathbf{r}'|} = \frac{q^2}{2} \int d^3r \int d^3r' \, \frac{\hat{n}(\mathbf{r})\hat{n}(\mathbf{r}') - \hat{n}(\mathbf{r})\delta(\mathbf{r} - \mathbf{r}')}{|\mathbf{r} - \mathbf{r}'|} \quad (A.49)$$

The operator character is only carried by the density operators $\hat{n}$ (in occupation number representation) and the term with the delta function subtracts out the interaction of a charge with itself (which is taken care of by $i < j$ in the other case). The expectation value of this operator $U = \langle \Phi|\hat{U}|\Phi \rangle$ involves the expectation value of a product of density operators $\langle \Phi|\hat{n}(\mathbf{r})\hat{n}(\mathbf{r}')|\Phi \rangle$, which in the Hartree term (A.47) is replaced by a product of expectation values, each of the form $n(\mathbf{r}) = \langle \Phi|\hat{n}(\mathbf{r})|\Phi \rangle$. This replacement amounts to a mean field approximation, which neglects quantum fluctuations. Therefore, the correlation energy accounts for the energy lowering due to quantum fluctuations, i.e. the 'correlation hole', which arises because electrons with unlike spins try to coordinate their movement to minimize their Coulomb energy. The other significant part of the correlation energy is due to the difference T_c between non-interacting and interacting kinetic energies.

A. Appendix

Nomenclature and properties regarding E_{xc}

- Since both exchange and correlation tend to keep electrons apart, the term 'electron hole' was coined, describing the region of reduced probability for encountering a second electron in the vicinity of a given reference electron.

- Both exchange and correlation energies give negative contributions, which leads to an upper bound of 0 for the energy correction, the lower one given by the Lieb-Oxford bound: $E_{xc} \geq -1.68 \int d^2 r \, n(\mathbf{r})^{4/3}$

- In the one electron limit $E_c(n^{(1)}) = 0$ and $E_x(n^{(1)}) = -U_H(n^{(1)})$, where $n^{(1)}$ is a one-electron density. Note: this is satisfied by HF, but not by standard LDA and GGA functionals.

The Kohn-Sham equations

To get the ground state energy, i.e. to minimize eqn. (A.45) with respect to the density, one cannot directly minimize with respect to n, since T_s is written as an orbital functional. Instead, Kohn and Sham proposed the following scheme for indirect minimization. To minimize E with respect to density means, we want:

$$0 = \frac{\delta E[n]}{\delta n(\mathbf{r})} = \frac{\delta T_s[n]}{\delta n(\mathbf{r})} + \frac{\delta V[n]}{\delta n(\mathbf{r})} + \frac{\delta U_H[n]}{\delta n(\mathbf{r})} + \frac{\delta E_{xc}[n]}{\delta n(\mathbf{r})} = \frac{\delta T_s[n]}{\delta n(\mathbf{r})} + v(\mathbf{r}) + v_H(\mathbf{r}) + v_{xc}(\mathbf{r}) \quad (A.50)$$

As a consequence of eqn. (A.44) $\frac{\delta V}{\delta n} = v(\mathbf{r})$, which is nothing other than the external potential due to the fixed nuclei, the lattice or a truly external field. The term $\frac{\delta U_H}{\delta n}$ yields the Hartree potential introduced in eqn. (A.40). The term $\frac{\delta E_{xc}}{\delta n}$ can only be calculated after an approximation has been chosen, but nevertheless we can call the result v_{xc}. Now consider this brillian idea: look at a system of non-interacting particles moving in the arbitrary potential $v_s(\mathbf{r})$. For this system the minimization condition is:

$$0 = \frac{\delta E_s[n]}{\delta n(\mathbf{r})} = \frac{\delta T_s[n]}{\delta n(\mathbf{r})} + \frac{\delta V_s[n]}{\delta n(\mathbf{r})} = \frac{\delta T_s[n]}{\delta n(\mathbf{r})} + v_s(\mathbf{r})$$

The density solving this Euler equation is $n_s(\mathbf{r})$. Comparison with eqn. (A.50) shows that both minimizations have the same solution $n_s(\mathbf{r}) = n(\mathbf{r})$, if we choose:

$$v_s(\mathbf{r}) = v(\mathbf{r}) + v_H(\mathbf{r}) + v_{xc}(\mathbf{r}) \quad (A.51)$$

Therefore we can calculate the density of an interacting many-body system in the potential $v(\mathbf{r})$, described by a many-body Schrödinger equation, by simply solving the equations of a noninteracting single-body system in the potential $v_s(\mathbf{r})$, given by:

$$\left[-\frac{\hbar^2}{2m}\nabla^2 + v_s(\mathbf{r})\right]\phi_i(\mathbf{r}) = \epsilon_i \phi_i(\mathbf{r}) \quad (A.52)$$

This yields orbitals that reproduce the density $n(\mathbf{r})$ of the original system via:

$$n(\mathbf{r}) = n_s(\mathbf{r}) = \sum_i^N f_i |\phi_i(\mathbf{r})|^2 , \qquad (A.53)$$

where f_i is the occupation of the ith orbital. Equations (A.51) to (A.53) are the celebrated Kohn-Sham equations. The problem of minimizing $E(n)$ (and originally of solving the corresponding many-body Schrödinger equation) is reduced to that of solving a noninteracting Schrödinger equation. The solution to this nonlinear problem is usually found by starting with an initial guess for $n(\mathbf{r})$, calculating the corresponding $v_s(\mathbf{r})$ and then solving the differential equation A.52 for the ϕ_i. From these a new density is calculated using eqn. (A.53) and the process is restarted until reasonable convergence is reached.

A.2.5. Construction of exchange functionals

Local density approximation (LDA)

The general idea of LDA is to take the known result for a homogeneous system and apply it locally to a non-homogeneous system. The exchange energy of a homogeneous system is known to be:

$$e_x^{hom}(n) = -\frac{3q^2}{4}\left(\frac{3}{\pi}\right)^{1/3} n^{4/3}$$

So in the LDA one takes:

$$E_x^{LDA}(n) = -\frac{3q^2}{4}\left(\frac{3}{\pi}\right)^{1/3} \int n(\mathbf{r})^{4/3} d^3r \qquad (A.54)$$

Expressions for $E_c^{LDA}(n)$ are parameterizations of accurate Quantum Monte Carlo (QMC) calculations for the electron liquid.

Gradient expansion approximation (GEA)

In this case one tries to systematically calculate gradient corrections of the form $|\nabla n(\mathbf{r})|$, $|\nabla n(\mathbf{r})|^2$, $\nabla^2 n(\mathbf{r})$ etc. to the LDA to take into account the rate of density variation in the system. The lowest order correction to the LDA exchange energy is given by the Weizsäcker term [26]:

$$E_x^{GEA(2)}(n) = E_x^{LDA}(n) - \frac{10q^2}{432\pi(3\pi^2)^{1/3}} \int d^3r \frac{|\nabla n(\mathbf{r})|^2}{n(\mathbf{r})^{4/3}} \qquad (A.55)$$

In practice, low-order gradient corrections almost never improve the LDA results and higher-order corrections are exceedingly difficult to calculate.

A. Appendix

Generalized gradient approximation (GGA)

It was a major breakthrough, when it was realized that instead of power-series-like systematic gradient expansions one could experiment with more general functions of $n(\mathbf{r})$ and $\nabla n(\mathbf{r})$, which need not proceed order by order. Such functional have the general form:

$$E_{xc}^{GGA}(n) = \int d^3r \, f(n(\mathbf{r}), \nabla n(\mathbf{r})) \tag{A.56}$$

The most popular GGAs are PBE (denoting the functional proposed in 1996 by Perdew, Burke and Ernzerhof) and BLYP (denoting the combination of the exchange functional by Becke and the correlation functional of Lee, Yang and Parr, both in 1988). GGAs give reliable results for chemical bonds, but mostly fail for dispersion interactions (van der Waals terms).

Meta-GGA

These functionals depend, in addition to the density and its derivatives, also on the Kohn-Sham kinetic-energy density $\tau(\mathbf{r})$.

$$\tau(\mathbf{r}) = \frac{\hbar^2}{2m} \sum_i |\nabla \phi_i(\mathbf{r})|^2 \tag{A.57}$$

The exchange-correlation energy is therefore written as $E_{xc}[n(\mathbf{r}), \nabla n(\mathbf{r}), \tau(\mathbf{r})]$. The additional degree of freedom is used to satisfy additional constraints on E_{xc}, such as a self-interaction correlation functional, etc. In recent tests, Meta-GGAs have performed favorably even when compared to the best GGAs.

B3LYP

One of the most popular functionals in quantum chemistry today is B3LYP [164], which combines the meta-GGA correlation functional LYP developed by Lee, Yang and Parr [165] with Becke's three-parameter hybrid functional B3 for exchange [166, 167]. The latter mixes a fraction of Hartree-Fock exchange into the DFT exchange functional. This mixing involves a certain amount of empiricism and optimization for selected classes of molecules.
The functional, which is implemented in *Gaussian* and hence used in our computation is given by:

$$E_{xc}^{B3LYP} = a_0 E_x^{exact} + (1-a_0) E_x^{LSDA} + a_x \Delta E_x^{B88} + E_c^{VWN-RPA} + a_c E_c^{LYP} \tag{A.58}$$

Here a_0, a_x and a_c are the three empirical parameters. They were determined by fitting to series of atomization energies, proton affinities and ionization potentials yielding $a_0 = 0.20$, $a_x = 0.72$ and $a_c = 0.81$ [168]. E_x^{exact} is the exchange energy calculated by Hartree-Fock-like methods. E_x^{LSDA} is the energy of a uniform free-electron gas (local spin density approximation). E^{B88}[169] satisfies the experimentally known asymptotic behavior of finite many-electron systems [170, 171]. $E_c^{VWN-RPA}$ is the local density approximation of the correlation energy introduced by Vosko, Wilk and Nusair[172] with parameters obtained by fitting to a uniform electron gas calculated in the random-phase approximation. Finally, E_c^{LYP} is the meta-GGA correlation functional of Lee, Yang and Parr [165].

A.3. Force field parameterization

A.3.1. [1]Benzothieno[3,2-b]benzothiophene (BTBT)

The atom types as well as the force constants were taken from the OPLS force field as far as possible by using analogies to thiazole, furan, benzene and indole. The atom type *CC* was introduced to denote the carbon connecting the phenyl rings with the side chains. The resulting atom labeling is shown in figure A.1. Equilibrium values for bond distances and angles were taken from x-ray data and quantum chemical geometry optimizations in case the differences to existing OPLS values exceeded 2 pm for bonds or one degree for angles. The planarity of the core was ensured by identical impropers. The side chain parameters had to be adjusted since they were modelled by united atoms. Ryckaert-Belleman dihedrals were used to describe the attachment of the united atom side chain to the core as well as the links within the side chain itself. An additional improper fixes the planarity of the first side chain atom with respect to the core. All parameters were calculated using *Gaussian* with B3LYP functional and 6-311G(d,p) level of accuracy in analogy to [93]. All self-calculated parameters as well as those equilibrium values taken from x-ray data to replace the OPLS entries are displayed in tables A.1 and A.2.

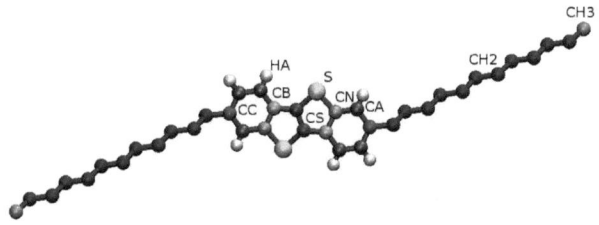

Fig. A.1.: Atom types used in the MD force field.

Bond	b_0	Angle	ϕ_0
CA-CA	0.138	CA-CA-CB	119.5
CA-CC	0.139	CA-CB-CN	118.8
CH2-CH2	0.153	CA-CB-CS	131.0
CH2-CH3	0.153	CA-CN-S	125.5

Tab. A.1.: Equilibrium bond lengths b_0 in nano meters and angles ϕ_0 in degrees.

Partial charges were calculated based on geometry optimization of a single molecule in vacuum using DFT with B3LYP/6-311G(d,p). The wave function was further optimized using higher level basis sets and finally charges were computed using CHELPG method [42]. Convergence of charge with basis set was checked starting from 3-21G(d) up to 6-311++(3df,3pd). An example of the convergence for the three different representative atoms, sulfur (S9), carbon (C4) and hydrogen (H3), is shown in figure A.2. On the side chains only the atom closest to the core has partial charge. The resulting charges are listed in table A.3. The dipole moment of the

A. Appendix

Dihedral	V_1	V_2	V_3	V_4	V_5
CH2-CH2	17.45	1.14	-27.2	0.0	0.0
CC-CH2	-0.03	9.44	0.49	-5.15	-0.47
Improper	ϕ_0	k_ϕ			
CA-CC-CA-CA	0.0	167.4			
CC-CA-CA-CH2	0.0	170.6			

Tab. A.2.: Parameters of the torsional potentials. Units are kJ mol^{-1} for Ryckaert-Belleman dihedrals and kJmol^{-1}rad^{-2} for impropers. The CA-CC-CA-CA improper constant is used for every improper in the core. For the dihedral definitions only the two central atoms of the dihedral are given.

optimized BTBT core is only 0.005 Debye, which justifies neglecting electrostatic disorder in the charge transport simulations.

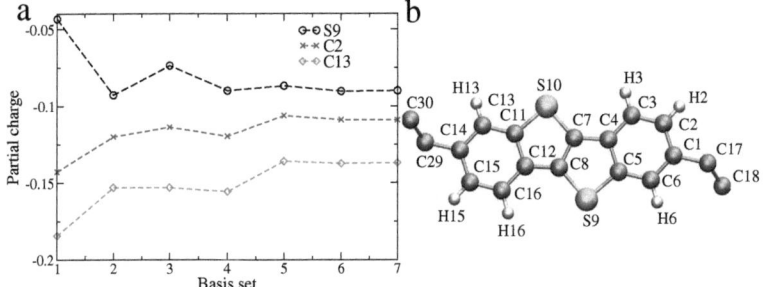

Fig. A.2.: (a) Convergence of partial charge with basis set. The basis sets (1-6) are: 3-21G(d), 6-31G(d), 6-311g(d,p), 6-311++G(d,p), 6-311++G(2d,2p) and 6-311++G(3df,3pd). Number seven corresponds to the final choice in the force field. (b) Atom labels for BTBT core and first side chain atoms.

A.3.2. Indolocarbazole

The force field parameters are based on those of carbazole (sec. 5.1), but in this case OPLS all atom parameters were used for the CH$_3$ side chains. The resulting atom types and labels are displayed in fig. A.3. The OPLS force field did not contain dihedral parameters for C_N-N_A-C_T-H_C, describing the attachment of the simple CH$_3$ side chains to the core, or for C_A-O_S-C_T-H_C and C_N-C_A-O_S-C_T, required to describe the rotation of the O-CH$_3$ side chains. Because hydrogen motion is not of major importance, we chose parameters already present in the OPLS force field. Two dihedral parameter sets were subject to consideration: C_A-O_S-C_T-H_C from ethers and C_T-C_T-C_T-H_C from hydrocarbons. They are compared in fig. A.4a. Both potentials correspond to the same equilibrium configurations with one of the three hydrogens at 0°, 120° or 240°. The only difference is the barrier height and we chose the stiffer one corresponding to C_A-O_S-C_T-H_C.

For C_N-C_A-O_S-C_T we compared the available OPLS-AA dihedral corresponding to C_A-C_A-O_S-C_T from anisole with parameters calculated based on B3LYP. The results of energy calculations with basis sets 6-31G(d), 6-311G(d,p) and 6-311++G(d,p) are shown in fig. A.4b. Next, we calculated the dihedral parameters by subtracting the force-field results without parameters for this

A.3 Force field parameterization

Label	Charge	Label	Charge
C1	0.045	C2	-0.109
C3	-0.280	C4	0.276
C5	-0.046	C6	-0.137
C7	-0.083	C8	-0.083
S9	-0.090	S10	-0.090
C11	-0.046	C12	0.276
C13	-0.137	C14	0.045
C15	-0.109	C16	-0.280
C17	0.010	C18	0.000
C29	0.010	C30	0.000

Tab. A.3.: Partial charges used for BTBT simulation given in multiples of the elementary charge e.

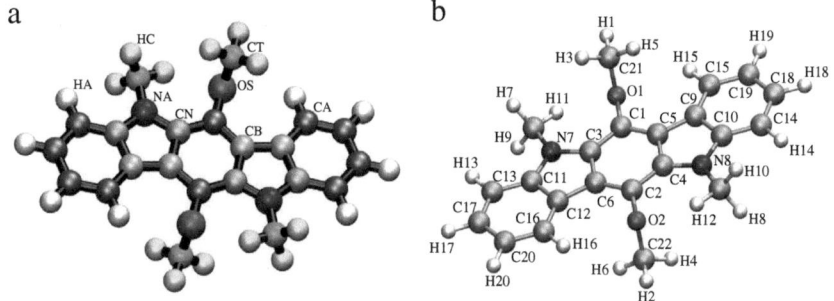

Fig. A.3.: (a) Atom types used in the Indolocarbazole force field. (b) Labeling of the atoms.

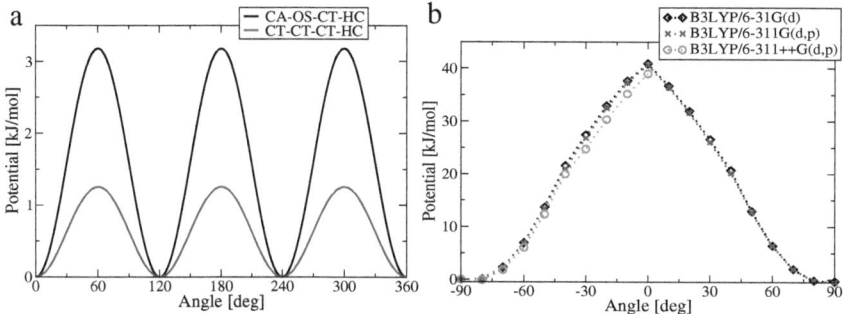

Fig. A.4.: (a) Comparison of the two dihedral parameter sets for H_C hydrogens in OPLS. (b) Results of B3LYP energy calculations for the C_N-C_A-O_S-C_T dihedral and different basis sets.

145

A. Appendix

dihedral from the DFT results and fitting to a Ryckaert-Belleman (RB) function. The geometries obtained by DFT calculation were optimized using the steepest decent method before calculating the corresponding energies with the force-field (see sec. 3.3.2). Thereafter we compared our dihedral parameters with those given in the OPLS force field. The significantly smaller barrier in case of OPLS-AA is due to the geometry of anisole, which allows the side chain to freely rotate, while this is sterically hindered in case of indolocarbazole by the CH_3 side chain and phenyl rings in the core. Lastly, we compared the total change in energy due to dihedral rotation from DFT calculations with the force-field based on our and the OPLS parameterizations in fig. A.5. The better agreement between B3LYP and force field energies based on the recalculated parameters at low values led us to choose them over the OPLS parameters for the following MD simulations. All parameters are presented in tab. A.4. In addition equilibrium angles were adjusted in case x-ray data and QC computations deviated too strongly from the OPLS force field values. Partial charges were calculated based on a B3LYP/6-311G(d,p) optimized geometry and converged with respect to basis set. Results are summarized in tab. A.5.

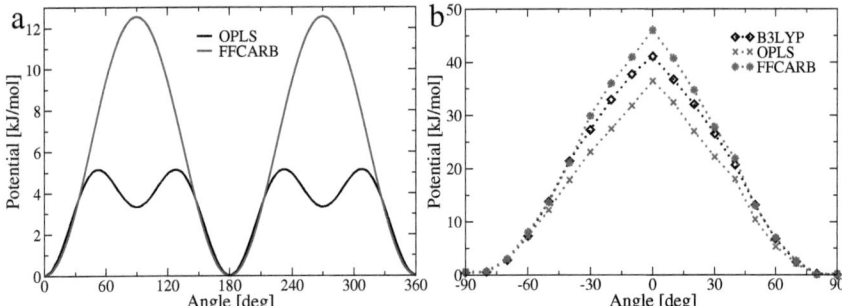

Fig. A.5.: (a) Self-calculated (FFCARB) and OPLS dihedral potentials. (b) Change in total energy upon rotation of dihedral as computed by B3LYP and the MD force-field using either FFCARB or OPLS dihedral parameters.

dihedral	V_1	V_2	V_3	V_4	V_5
$C_N - C_A - O_S - C_T$ (OPLS)	0.0	-12.552	0.0	0.0	0.0
$C_N - C_A - O_S - C_T$ (FFCARB)	0.0	9.702	0.0	-13.016	0.0
$C_A - C_A - O_S - C_T$	0.0	-12.552	0.0	0.0	0.0
$C_A - O_S - C_T - H_C$	4.770	0.0	-6.360	0.0	0.0
$C_N - N_A - C_T - H_C$	4.770	0.0	-6.360	0.0	0.0
$C_T - C_T - C_T - H_C$	1.883	0.0	-2.510	0.0	0.0

Tab. A.4.: Coefficients of the Ryckaert Belleman function (eqn. 5.1) for the side chain dihedrals in kJ/mol. FFCARB refers to our calculated force field parameters.

A.3.3. Benzo[1,2-b:4,5-b']bis[b]benzothiophene (BBBT)

The force field used is almost identical to that of BTBT except that some atom types had to be changed to incorporate the central phenyl ring. Again, side chains were treated using the

A.3 Force field parameterization

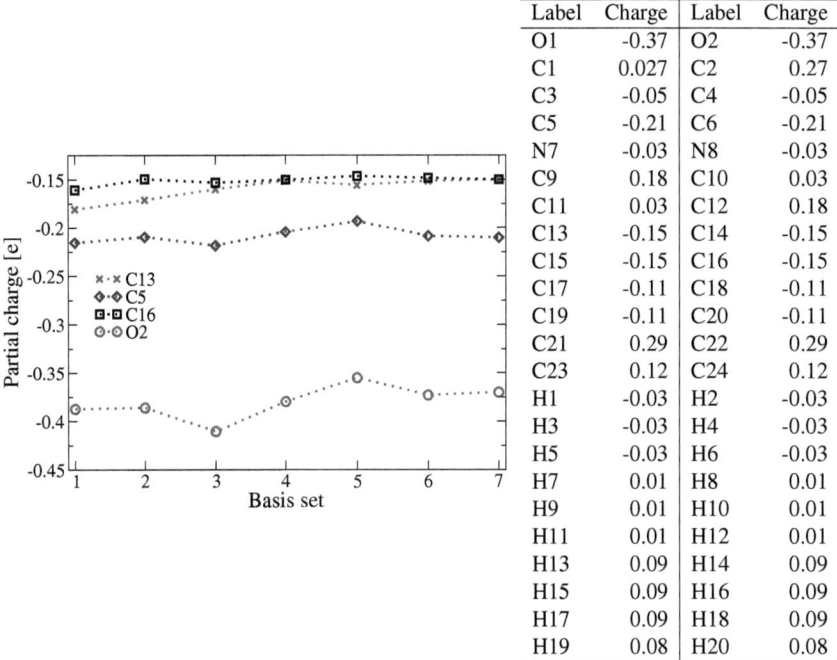

Label	Charge	Label	Charge
O1	-0.37	O2	-0.37
C1	0.027	C2	0.27
C3	-0.05	C4	-0.05
C5	-0.21	C6	-0.21
N7	-0.03	N8	-0.03
C9	0.18	C10	0.03
C11	0.03	C12	0.18
C13	-0.15	C14	-0.15
C15	-0.15	C16	-0.15
C17	-0.11	C18	-0.11
C19	-0.11	C20	-0.11
C21	0.29	C22	0.29
C23	0.12	C24	0.12
H1	-0.03	H2	-0.03
H3	-0.03	H4	-0.03
H5	-0.03	H6	-0.03
H7	0.01	H8	0.01
H9	0.01	H10	0.01
H11	0.01	H12	0.01
H13	0.09	H14	0.09
H15	0.09	H16	0.09
H17	0.09	H18	0.09
H19	0.08	H20	0.08

Tab. A.5.: (a) Charge convergence with respect to basis set. The numbers one through six refer to 3-21G(d), 6-31G(d), 6-311G(d,p), 6-311++G(d,p), 6-311++G(2d,2p) and 6-311G(2df,2pd), respectively. Number seven is the final choice. The table summarizes the final partial charges for indolocarbazole with CH_3 side chains. For labeling refer to fig. A.3.

A. Appendix

united atom approach and equilibrium angles were adjusted in case the QC optimized geometry or the x-ray data deviated too strongly from the initial OPLS values. Atom types and labels are shown in fig. A.6. Partial charges were calculated based on a B3LYP/6-311G(d,p) optimized geometry and convergence with respect to basis set was checked up to 6-311++G(3df,3pd) for both compounds. Except for the first, all side chain atoms were considered to have neutral charge.

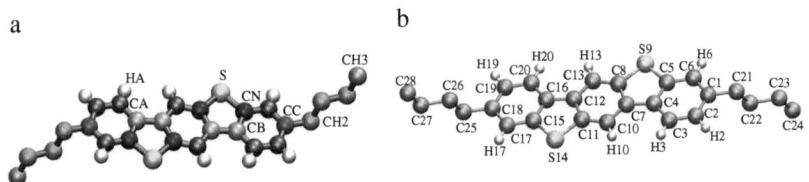

Fig. A.6.: (a) Atom types used in the BBBT force field. (b) Labeling of the atoms. In case of BBBT without side chains the atom type C_C does not exist and the side chain is replace by an H_A atom.

BBBT (1)				BBBT (2)			
Label	Charge	Label	Charge	Label	Charge	Label	Charge
C1	-0.158	C2	-0.006	C1	0.099	C2	-0.149
C3	-0.176	C4	0.106	C3	-0.127	C4	0.031
C5	0.045	C6	-0.074	C5	0.094	C6	-0.224
C7	-0.015	C8	0.083	C7	0.027	C8	0.088
S9	-0.150	C10	-0.183	S9	-0.162	C10	-0.210
C11	0.083	C12	-0.015	C11	0.088	C12	0.027
C13	-0.183	S14	-0.150	C13	-0.210	S14	-0.162
C15	0.045	C16	0.106	C15	0.094	C16	0.031
C17	-0.074	C18	-0.158	C17	-0.224	C18	0.099
C19	-0.006	C20	-0.176	C19	-0.149	C20	-0.127
H1	0.087	H2	0.067	C21	-0.007	C25	-0.007
H3	0.80	H6	0.120	H2	0.109	H3	0.077
H10	0.174	H13	0.174	H6	0.176	H10	0.178
H17	0.120	H18	0.087	H13	0.178	H17	0.176
H19	0.067	H20	0.080	H19	0.109	H20	0.077

Tab. A.6.: Partial charges used in the force field for BBBT (1) without and (2) with C_4H_9 side chains. For labeling refer to fig. A.6.

Bibliography

[1] Report on the basic energy sciences. Department of Energy, http://www.energy.gov/.

[2] Kirkpatrick, J. *Int. J. Quant. Chem.*, **2008**, *108*, 51.

[3] Heeger, A. J. *Rev. Mod. Phys.*, Sep 2001, *73*(3), 681–700.

[4] Deibel, C. Current-voltage characteristics of organic solar cells, 2008. http://blog.disorderedmatter.eu/tag/organic-solar-cells/page/2/.

[5] Li, J.-F.; Chen, S.-F.; Su, S.-H.; Hwang, K.-S.; Yokoyama, M. *Applied Surface Science*, **2006**, *253*(5), 2522 – 2524.

[6] Kumar, S.; Ram, S. K.; Shukla, V.; Gupta, G. Organic semicoductors (small molecules). http://sanjayk.ram.googlepages.com/organic.

[7] Lilienfeld, J. E. Method and apparatus for controlling electric currents, 1930. United States Patent US 1,745,175, 28.01.1930.

[8] Schmieder, H. Ofet: Organischer feldeffekttransistor. http://www.pointtranslations.com/links_informationen/-technologie/ofet_oled_organischer-feldeffekttransistor/.

[9] Katz, H. E. *Journal of Materials Chemistry*, **1997**, *7*, 369–376.

[10] Horowitz, G. *Advanced Materials*, **1998**, *10*(5), 365–377.

[11] Im, C.; Bässler, H.; Rost, H.; Hörhold, H. H. *The Journal of Chemical Physics*, **2000**, *113*(9), 3802–3807.

[12] Poplavskyy, D.; Nelson, J. *Journal of Applied Physics*, **2003**, *93*(1), 341–346.

[13] Markham, J. P. J.; Anthopoulos, T. D.; Samuel, I. D. W.; Richards, G. J.; Burn, P. L.; Im, C.; Bässler, H. *Applied Physics Letters*, **2002**, *81*(17), 3266–3268.

[14] Laquai, F.; Wegner, G.; Im, C.; Bässler, H.; Heun, S. *Journal of Applied Physics*, **2006**, *99*(2), 023712.

[15] Laquai, F.; Wegner, G.; Im, C.; Bässler, H.; Heun, S. *Journal of Applied Physics*, **2006**, *99*(3), 033710.

[16] Infelta, P. P.; de Haas, M. P.; Warman, J. M. *Radiation Physics and Chemistry*, **1977**, *10*, 353–365.

[17] Piris, J. *Optoelectronic Properties of Discotic Materials for Device Applications*. PhD thesis, TU Delft, 2004.

[18] Schouten, P. G.; Warman, J. M.; de Haas, M. P. *The Journal of Physical Chemistry*, **1993**, *97*(38), 9863–9870.

[19] Warman, J. M.; Schouten, P. G.; Gelinck, G. H.; de Haas, M. P. *Chemical Physics*, **1996**, *212*(1), 183 – 192.

[20] van de Craats, A. M.; Warman, J. M.; de Haas, M. P.; Adam, D.; Simmerer, J.; Haarer, D.; Schuhmacher, P. *Advanced Materials*, **1996**, *8*(10), 823–826.

[21] Percec, V.; Glodde, M.; Bera, T. K.; Miura, Y.; Shiyanovskaya, I.; Singer, K. D.; Balagurusamy, V. S. K.; Heiney, P. A.; Schnell, I.; Rapp, A.; Spiess, H.-W.; Hudson, S. D.; Duan, H. *Nature*, **2002**, *417*, 384–387.

[22] Feng, X.; Marcon, V.; Pisula, W.; Hansen, M.; Kirkpatrick, J.; Grozema, F.; Andrienko, D.; Kremer, K.; Müllen, K. *Nature Materials*, **2009**, *8*, 421–426.

[23] Warman, J. M.; De Haas, M. P.; Van Hovell tot Westerflier, S. W. F. M.; Binsma, J. J. M.; Kolar, Z. I. *The Journal of Physical Chemistry*, **1989**, *93*(15), 5895–5899.

[24] Coropceanu, V.; Cornil, J.; da Silva Filho, D. A.; Olivier, Y.; Silbey, R.; Bredas, J.-L. *Chem. Rev.*, **2007**, *107*(4), 926–952.

[25] Cramer, C. J., Ed. *Essentials of Computational Chemistry - Theories and Models*. Wiley-VCH, Inc., 2002.

[26] Capelle, K. *Condensed Matter*, **2002**.

[27] Kohn, W. *Rev. Mod. Phys.*, Oct 1999, *71*(5), 1253–1266.

[28] Jorgensen, W.; Madura, J.; Swenson, C. *Journal of the American Chemical Society*, **1984**, *106*, 6638–6646.

[29] Jorgensen, W. L.; Tirado-Rives, J. *Journal of the American Chemical Society*, **1988**, *110*(6), 1657–1666.

[30] Jorgensen, W.; Severance, D. *Journal of the American Chemical Society*, **1990**, *112*, 4768–4774.

[31] Jorgensen, W.; Laird, E.; Nguyen, T.; Tirado Rives, J. *Journal of Computational Chemistry*, **1993**, *14*, 206–215.

[32] Jorgensen, W. L.; Maxwell, D. S.; Tirado-Rives, J. *Journal of the American Chemical Society*, **1996**, *118*(45), 11225–11236.

[33] Jorgensen, W. L.; Tirado-Rives, J. *Proceedings of the National Academy of Sciences of the United States of America*, **2005**, *102*(19), 6665–6670.

[34] Weiner, S. J.; Kollman, P. A.; Case, D. A.; Singh, U. C.; Ghio, C.; Alagona, G.; Profeta, S.; Weiner, P. *Journal of the American Chemical Society*, **1984**, *106*(3), 765–784.

[35] Cornell, W. D.; Cieplak, P.; Bayly, C. I.; Kollmann, P. A. *Journal of the American Chemical Society*, **1993**, *115*(21), 9620–9631.

[36] Cornell, W. D.; Cieplak, P.; Bayly, C. I.; Gould, I. R.; Merz, K. M.; Ferguson, D. M.; Spellmeyer, D. C.; Fox, T.; Caldwell, J. W.; Kollman, P. A. *Journal of the American Chemical Society*, **1995**, *117*(19), 5179–5197.

[37] Duan, Y.; Wu, C.; Chowdhury, S.; Lee, M.; Xiong, G.; Zhang, W.; Yang, R.; Cieplak, P.; Luo, R.; Lee, T.; Caldwell, J.; Wang, J.; Kollmann, P. *Journal of Computational Chemistry*, **2003**, *24*(16), 1999–2012.

[38] Allinger, N. L.; Yuh, Y. H.; Lii, J. H. *Journal of the American Chemical Society*, **1989**, *111*(23), 8551–8566.

[39] Allinger, N. L.; Chen, K.; Lee, J.-H. *Journal of Computational Chemistry*, **1996**, *17*(5-6), 642–668.

[40] Mackerell, A. D. *Journal of Computational Chemistry*, **2004**, *25*(13), 1584–1604.

[41] PumMa. Theory: Potentials. http://www.pumma.nl/index.php/Theory/Potentials.

[42] Breneman, C. B.; Wiberg, K. E. *J. Comp. Chem.*, **1990**, *11*(3), 361–373.

[43] Bayly, C. I.; Cieplak, P.; Cornell, W.; Kollman, P. A. *The Journal of Physical Chemistry*, **1993**, *97*(40), 10269–10280.

[44] Verlet, L. *Phys. Rev.*, Jul 1967, *159*(1), 98.

[45] Hockney, R. W.; Goel, S. P.; Eastwood, J. W. *Journal of Computational Physics*, **1974**, *14*(2), 148 – 158.

[46] D. Van der Spoel, B. H. E. Lindahl, Ed. *Gromacs User Manual Version 4.0*. www.gromacs.org, 2008.

[47] Hünenberger, P. H. *Advances in Polymer Science*, **2005**, *173*, 105–149.

[48] Berendsen, H. J. C.; Postma, J. P. M.; van Gunsteren, W. F.; DiNola, A.; Haak, J. R. *The Journal of Chemical Physics*, **1984**, *81*(8), 3684–3690.

[49] Nosé, S. *Molecular Physics*, **1984**, *52*(2), 255–268.

[50] Hoover, W. G. *Phys. Rev. A*, Mar 1985, *31*(3), 1695–1697.

[51] Bussi, G.; Donadio, D.; Parrinello, M. *The Journal of Chemical Physics*, **2007**, *126*(1), 014101.

[52] Schneider, T.; Stoll, E. *Phys. Rev. B*, Feb 1978, *17*(3), 1302–1322.

[53] Soddemann, T.; Dünweg, B.; Kremer, K. *Phys. Rev. E*, Oct 2003, *68*(4), 046702.

[54] Espanol, P.; Warren, P. *Europhys. Lett.*, **1995**, *30*, 191–196.

[55] Parrinello, M.; Rahman, A. *Journal of Applied Physics*, **1981**, *52*(12), 7182–7190.

[56] Voter, A. F., Ed. *Radiation effects in solids*. Springer, NATO Publishing Unit, Dordrecht, 2005.

[57] Hammersley, J. M.; Handscomb, D. C., Eds. *Monte Carlo Methods*. Methuen & Co Ltd., London, 1965.

[58] Metropolis, N.; Rosenbluth, A. W.; Rosenbluth, M. N.; Teller, A. H.; Teller, E. *The Journal of Chemical Physics*, **1953**, *21*(6), 1087–1092.

[59] Beeler, J. R. *Phys. Rev.*, Oct 1966, *150*(2), 470–487.

[60] Bortz, A. B.; Kalos, M. H.; Lebowitz, J. L. *Journal of Computational Physics*, **1975**, *17*(1), 10 – 18.

[61] Luttinger, J. M.; Kohn, W. *Phys. Rev.*, Feb 1955, *97*(4), 869–883.

[62] Kohn, W.; Luttinger, J. M. *Phys. Rev.*, May 1955, *98*(4), 915–922.

[63] Hutchison, G. R.; Zhao, Y.-J.; Delley, B.; Freeman, A. J.; Ratner, M. A.; Marks, T. J. *Phys. Rev. B*, Jul 2003, *68*(3), 035204.

[64] Kim, E.-G.; Coropceanu, V.; Gruhn, N.; Sánchez-Carrera, R.; Snoeberger, R.; Matzger, A.; Brédas, J.-L. *Journal of the American Chemical Society*, **2007**, *129*(43), 13072–13081.

[65] Troisi, A.; Orlandi, G. *The Journal of Physical Chemistry B*, **2005**, *109*(5), 1849–1856.

[66] Troisi, A. *Advances in Polymer Science*, **2009**.

[67] Holstein, T. *Annals of Physics*, **2000**, *281*(1-2), 706–724.

[68] Holstein, T. *Annals of Physics*, **2000**, *281*(1-2), 725–773.

[69] Munn, R. W.; Silbey, R. *The Journal of Chemical Physics*, **1985**, *83*(4), 1843–1853.

[70] Munn, R. W.; Silbey, R. *The Journal of Chemical Physics*, **1985**, *83*(4), 1854–1864.

[71] Hannewald, K.; Bobbert, P. A. *Phys. Rev. B*, Feb 2004, *69*(7), 075212.

[72] Hannewald, K.; Bobbert, P. A. *Applied Physics Letters*, **2004**, *85*(9), 1535–1537.

[73] Moorthy, J. N.; Venkatakrishnan, P.; Savitha, G.; Weiss, R. G. *Photochemical & Photobiological Sciences*, **2006**, *5*, 903–913.

[74] Troisi, A.; Orlandi, G. *J. Phys. Chem.*, **2006**, *110*, 4065–4070.

[75] Troisi, A. *Advanced Materials*, **2007**, *19*(15), 2000–2004.

[76] Malagoli, M.; Coropceanu, V.; da Silva Filho, D. A.; Brédas, J. L. *The Journal of Chemical Physics*, **2004**, *120*(16), 7490–7496.

[77] D. A. da Silva Filho, J.-L. B. E.-G. Kim. *Advanced Materials*, **2005**, *17*(8), 1072–1076.

[78] Jortner, J. *The Journal of Chemical Physics*, **1976**, *64*(12), 4860–4867.

[79] Miller, A.; Abrahams, E. *Phys. Rev.*, Nov 1960, *120*(3), 745–755.

[80] Bässler, H. *Physica Status Solidi B*, **1993**, *175*(1), 15–56.

[81] Marcus, R. A. *The Journal of Chemical Physics*, **1956**, *24*(5), 966–978.

[82] Marcus, R. A. *Rev. Mod. Phys.*, **1993**, *65*(3), 599–610.

[83] Marcus, R. A. *The Journal of Chemical Physics*, **1957**, *26*(4), 867–871.

[84] Marcus, R. A. *The Journal of Chemical Physics*, **1957**, *26*(4), 872–877.

[85] Marcus, R. A. *Discussions of the Faraday Society*, **1960**, *29*, 21.

[86] Brédas, J.-L.; Beljonne, D.; Coropceanu, V.; Cornil, J. *Chemical Reviews*, **2004**, *104*(11), 4971–5004.

[87] Freed, K. F.; Jortner, J. *The Journal of Chemical Physics*, **1970**, *52*(12), 6272–6291.

[88] Kestner, N. R.; Logan, J.; Jortner, J. *The Journal of Physical Chemistry*, **1974**, *78*(21), 2148–2166.

[89] Miller, J.; Calcaterra, L.; Closs, G. *Journal of the American Chemical Society*, **1984**, *106*(10), 3047–3049.

[90] Jankowiak, R.; Rockwitz, K.; Bässler, H. *The Journal of Physical Chemistry*, **1983**, *87*(4), 552–557.

[91] Coehoorn, R.; Pasveer, W. F.; Bobbert, P. A.; Michels, M. A. J. *Phys. Rev. B*, Oct 2005, *72*(15), 155206.

[92] Kirkpatrick, J.; Marcon, V.; Nelson, J.; Kremer, K.; Andrienko, D. *Phys. Rev. Lett.*, **2007**, *98*, 227402.

[93] Marcon, V.; Vehoff, T.; Kirkpatrick, J.; Jeong, C.; Yoon, D. Y.; Kremer, K.; Andrienko, D. *J. Chem. Phys.*, **2008**, *129*, 094505.

[94] Kirkpatrick, J.; Marcon, V.; Kremer, K.; Nelson, J.; Andrienko, D. *J. Chem. Phys.*, **2008**, *129*, 094506.

[95] Rühle, V.; Junghans, C.; Lukyanov, A.; Kremer, K.; Andrienko, D. *Journal of Chemical Theory and Computation*, **2009**, *5*(12).

[96] Newton, M. D. *Chemical Reviews*, **1991**, *91*(5), 767–792.

[97] Doi, K.; Yoshida, K.; Nakano, H.; Tachibana, A.; Tanabe, T.; Kojima, Y.; Okazaki, K. *Journal of Applied Physics*, **2005**, *98*(11), 113709.

[98] Sancho-García, J. C.; Pérez-Jiménez, A. J. *The Journal of Chemical Physics*, **2008**, *129*(2), 024103.

[99] Lipparini, F.; Mennucci, B. *The Journal of Chemical Physics*, **2007**, *127*(14), 144706.

[100] Vehoff, T.; Kirkpatrick, J.; Kremer, K.; Andrienko, D. *physica status solidi (b)*, **2008**, *245*(5), 839–843.

[101] Sariciftci, N. S.; Wudla, F.; Heeger, A. J.; Maggini, M.; Scorrano, G.; Prato, M.; Bourassa, J.; Ford, P. C. *Chemical Physics Letters*, **1995**, *247*(4), 510–514.

[102] Wiberg, J.; Guo, L.; Pettersson, K.; Nilsson, D.; Ljungdahl, T.; Martensson, J.; Albinsson, B. *Journal of the American Chemical Society*, **2007**, *129*(1), 155–163.

[103] Jung, S.-H.; Pisula, W.; Rouhanipour, A.; Räder, H. J.; Jacob, J.; Müllen, K. *Angewandte Chemie International Edition*, **2006**, *45*(28), 4685–4690.

[104] Castex, M. C.; Olivero, C.; Pichler, G.; Adss, D.; Cloutet, E.; Siove, A. *Synthetic Metals*, **2001**, *122*(1), 59 – 61. Proceedings of the E-MRS 2000 Spring Meeting, Symposium I.

[105] Li, J.; Dierschke, F.; Wu, J.; Grimsdale, A. C.; Müllen, K. *Journal of Materials Chemistry*, **2006**, *16*, 96–100.

[106] Schmaltz, B.; Rouhanipour, A.; Räder, H. J.; Pisula, W.; Müllen, K. *Angewandte Chemie International Edition Engl.*, **2009**, *48*(4), 720–724.

[107] Sugerman, G. Recovery of anthracene and carbazole, 1971. United States Patent US 3,624,174.

[108] Grazulevicius, J. V.; Strohriegl, P.; Pielichowski, J.; Pielichowski, K. *Progress in Polymer Science*, **2003**, *28*(9), 1297 – 1353.

[109] Wagner, J.; Pielichowski, J.; Hinsch, A.; Pielichowski, K.; Bogdal, D.; Pajda, M.; Kurek, S. S.; Burczyk, A. *Synthetic Metals*, **2004**, *146*(2), 159 – 165.

[110] Lee, J.-H.; Woo, H.-S.; Kim, T.-W.; Park, J.-W. *Optical Materials*, **2003**, *21*(1-3), 225 – 229.

[111] Cheung, D. L.; McMahon, D. P.; Troisi, A. *J. Phys. Chem. B*, **2009**, *113*(28), 9393–9401.

[112] Jorgensen, W. in *The Encyclopedia of Computational Chemistry*. volume 3. p 1986. 1998.

[113] Andrienko, D.; Marcon, V.; Kremer, K. *J. Chem. Phys.*, **2006**, *125*, 124902.

[114] Frisch, M. J. Gaussian 03. Gaussian, Inc., Wallingford, CT, 2004.

[115] Essmann, U.; Perera, L.; Berkowitz, M. L.; Darden, T.; Lee, H.; Pedersen, L. G. *The Journal of Chemical Physics*, **1995**, *103*(19), 8577–8593.

[116] Holstein, T. *Annals of Physics*, **1959**, *8*(3), 325 – 342.

[117] Zoppi, L.; Calzolari, A.; Ruini, A.; Ferretti, A.; Caldas, M. *Phys. Rev. B*, **2008**, *78*.

[118] Paul. F. Barbara, T. J. M.; Ratner, M. A. *The Journal of Physical Chemistry*, **1996**, *100*(31), 13148–13168.

[119] Rühle, V.; Kirkpatrick, J.; Andrienko, D. *The Journal of Chemical Physics*, **2010**, *132*.

[120] Dimitrakopoulos, C. D.; Malenfant, P. R. L.

[121] Stingelin-Stutzmann, N.; Smits, E.; Wondergem, H.; Tanase, C.; Blom, P.; Smith, P.; de Leeuw, D.

[122] Takimiya, K.; Ebata, H.; Sakamoto, K.; Izawa, T.; Otsubo, T.; Kunugi, Y. *Journal of the American Chemical Society*, **2006**, *128*(39), 12604–12605.

[123] Ebata, H.; Izawa, T.; Miyazaki, E.; Takimiya, K.; Ikeda, M.; Kuwabara, H.; Yui, T. *J. Am. Chem. Soc.*, **2007**, *129*(51), 15732–15733.

[124] Vehoff, T.; Troisi, A.; Andrienko, D. *The Journal of Physical Chemistry C*, **2010**. accepted.

[125] Hess, B.; Kutzner, C.; van der Spoel, D.; Lindahl, E. *J. Chem. Theory Comput.*, **2008**, *4*(435).

[126] Lemaur, V.; da Silva Filho, D. A.; Coropceanu, V.; Lehmann, M.; Geerts, Y.; Piris, J.; Debije, M. G.; van de Craats, A. M.; Senthilkumar, K.; Siebbeles, L. D. A.; Warman, J. M.; Bredas, J.-L.; Cornil, J. *J. Am. Chem. Soc.*, **2004**, *126*(10), 3271–3279.

[127] Ness, H.; Fisher, A. J. *Phys. Rev. Lett.*, **1999**, *83*(2), 452–455.

[128] Nan, G.; Wang, L.; Yang, X.; Shuai, Z.; Zhao, Y. *J. Chem. Phys.*, **2009**, *130*(2), 024704.

[129] Troisi, A.; Nitzan, A.; Ratner, M. A. *J. Chem. Phys.*, **2003**, *119*(12), 5782.

[130] Herwig, P. T.; Müllen, K. *Advanced Materials*, **1999**, *11*(6), 480–483.

[131] Murphy, A. R.; Frchet, J. M. J.; Chang, P.; Lee, J.; Subramanian, V. *Journal of the American Chemical Society*, **2004**, *126*(6), 1596–1597.

[132] Haemori, M.; Yamaguchi, J.; Yaginuma, S.; Itaka, K.; Koinuma, H. *Jpn. J. Appl. Phys.*, **2005**, *44*, 3740 – 3742.

[133] Park, S. K.; Jackson, T. N.; Anthony, J. E.; Mourey, D. A. *Applied Physics Letters*, **2007**, *91*(6), 063514.

[134] Podzorov, V.; Menard, E.; Borissov, A.; Kiryukhin, V.; Rogers, J. A.; Gershenson, M. E. *Phys. Rev. Lett.*, Aug 2004, *93*(8), 086602.

[135] Sundar, V.; Zaumseil, J.; Podzorov, V.; Menard, E.; Willet, R.; Someya, T.; Gershenson, M.; Rogers, J. *Science*, **2004**, *303*, 1644.

[136] Podzorov, V.; Menard, E.; Rogers, J. A.; Gershenson, M. E. *Phys. Rev. Lett.*, Nov 2005, *95*(22), 226601.

[137] Ostroverkhova, O.; Cooke, D. G.; Hegmann, F. A.; Anthony, J. E.; Podzorov, V.; Gershenson, M. E.; Jurchescu, O. D.; Palstra, T. T. M. *Applied Physics Letters*, **2006**, *88*(16), 162101.

[138] Wrobel, N. *Synthese und photophysikalische Eigenschaften von Indolocarbazolen und hoeheren Analoga*. PhD thesis, Johannes Gutenberg Universität Mainz, 2008.

[139] Gao, P.; Beckmann, D.; Tsao, H. N.; Feng, X.; Enkelmann, V.; Pisula, W.; Müllen, K. *Chemical Communications*, **2008**, pp 1548–1550.

[140] Kloc, C.; Tan, K.; Toh, M.; Zhang, K.; Xu, Y. *Appl. Phys. A*, **2008**, *95*(1), 219–224.

[141] Aleshin, A. N.; Lee, J. Y.; Chu, S. W.; Kim, J. S.; Park, Y. W. *Applied Physics Letters*, **2004**, *84*(26), 5383–5385.

[142] Pernstich, K. P.; Rössner, B.; Batlogg, B.

[143] Tinker 5.1. http://dasher.wustl.edu/tinker/.

[144] Zeng, X.; Zhang, D.; Duan, L.; Wang, L.; Dong, G.; Qiu, Y. *Applied Surface Science*, **2007**, *253*(14), 6047 – 6051.

[145] Zhang, M.; Tsao, H. N.; Pisula, W.; Yang, C.; Mishra, A. K.; Müllen, K.

[146] Tsao, H. N.; Cho, D.; Andreasen, J. W.; Rouhanipour, A.; Breiby, D. W.; Pisula, W.; Müllen, K.

[147] de Craats, A. V.; Warman, J. M.; Müllen, K.; Geerts, Y.; Brand, J. *Advanced Materials*, **1998**, *10*(36).

[148] Hartree, D. R. *Mathematical Proceedings of the Cambridge Philosophical Society*, **1928**, *24*(01), 89–110.

[149] Hartree, D. R. *Mathematical Proceedings of the Cambridge Philosophical Society*, **1928**, *24*(01), 111–132.

[150] Hartree, D. R. *Mathematical Proceedings of the Cambridge Philosophical Society*, **1928**, *24*(03), 426–437.

[151] Slater, J. C. *Phys. Rev.*, Jan 1930, *35*(2), 210–211.

[152] Roothaan, C. C. J. *Rev. Mod. Phys.*, Apr 1951, *23*(2), 69–89.

[153] Zerner, M. C. *Theor. Chem. Acc.*, **2000**, *103*, 217–218.

[154] Pople, J. A.; Santry, D. P.; Segal, G. A. *The Journal of Chemical Physics*, **1965**, *43*(10), S129–S135.

[155] Pople, J. A.; Segal, G. A. *The Journal of Chemical Physics*, **1965**, *43*(10), S136–S151.

[156] Pariser, R.; Parr, R. G. *The Journal of Chemical Physics*, **1953**, *21*(5), 767–776.

[157] Pople, J. A.; Beveridge, D. L.; Dobosh, P. A. *The Journal of Chemical Physics*, **1967**, *47*(6), 2026–2033.

[158] Ridley, J.; Zerner, M. *Theoretica Chimica Acta*, **1973**, *32*(2), 111–134.

[159] Hehre, W. J.; Stewart, R. F.; Pople, J. A. *The Journal of Chemical Physics*, **1969**, *51*(6), 2657–2664.

[160] Thom H. Dunning, J. *The Journal of Chemical Physics*, **1989**, *90*(2), 1007–1023.

[161] Hohenberg, P.; Kohn, W. *Physical Review*, **1964**, *136*(3B), 864–871.

[162] Levy, M. *Proceedings of the National Academy of Sciences of the United States of America*, **1979**, *76*(12), 6062–6065.

[163] Levy, M. *Phys. Rev. A*, Sep 1982, *26*(3), 1200–1208.

[164] Stephens, P. J.; Devlin, F. J.; Chabalowski, C. F.; Frisch, M. J. *The Journal of Physical Chemistry*, **1994**, *98*(45), 11623–11627.

[165] Lee, C.; Yang, W.; Parr, R. G. *Phys. Rev. B*, Jan 1988, *37*(2), 785–789.

[166] Becke, A. D. *The Journal of Chemical Physics*, **1997**, *107*(20), 8554–8560.

[167] Becke, A. D. *Journal of Computational Chemistry*, **1997**, *20*(1), 63–69.

[168] Becke, A. D. *The Journal of Chemical Physics*, **1993**, *98*(7), 5648–5652.

[169] Becke, A. D. *Phys. Rev. A*, Sep 1988, *38*(6), 3098–3100.

[170] Gunnarsson, O.; Lundqvist, B. I. *Phys. Rev. B*, May 1976, *13*(10), 4274–4298.

[171] March, N. H. *Phys. Rev. A*, Nov 1987, *36*(10), 5077–5078.

[172] Vosko, S. H.; Wilk, L.; Nusair, M. *Canadian Journal of Physics*, **1980**, *58*, 1200–1211.

Acknowledgments

First and foremost I would like to thank Kurt Kremer for allowing me to join his group at the Max Planck Institute for Polymer Research and supervising my PhD work. It enabled me to work with excellent colleagues, attend conferences and workshops around the globe and gain some very good friends. I am also grateful to Denis Andrienko for supervising the work as my project leader, giving countless invaluable insights and spending quite a few enjoyable evenings with me all over the world. Do Yeung Yoon is gratefully acknowledged for the hospitality I experienced while staying in his group at Seoul National University as well as for valuable scientific discussions. Thanks also go to Alessandro Troisi for teaching me how to use his semi-classical dynamics code, allowing me to use the code in my thesis and being available for scientific discussions. I am also grateful to Rudolf Zentel for the coordination of the IRTG program, which allowed me to spend more than half a year in Korea and attend international conferences. I thank Wolfgang Paul for co-supervising my work and engulfing in helpful scientific discussions.

Lots of thanks go to Valentina Marcon for being there to answer lots of basic questions regarding molecular dynamics and for many more or less scientific but always entertaining discussions. Victor Rühle is very gratefully acknowledged for his infinite patience in answering programming questions, insightful discussions on charge transport and for joining me on many different trips and evenings. I also thank James Kirkpatrick for scientific discussions and help with the molecular orbital overlap code. Björn Baumeier is acknowledged for helping make the thesis more readable and understandable. Many thanks go to Dominik Fritz my office mate who made every day life at the institute very pleasant and was always available for hints, comments, proofreading, icecream, coffee and bar visits. I thank Christoph Junghans for tips regarding scripting, scientific discussions and introducing me to climbing. Berk Hess and Alessandra Villa are acknowledged for their excellent support regarding Gromacs and general programming issues. Alexander Lukyanov is acknowledged for helpful discussions on charge transport matters and accompanying me on many sight-seeing trips. I also wish to thank Doris Kirsch for being kind, friendly and helpful throughout my work. Thanks go to Marco Köhl for computer support and many sauna and movie evenings. I am grateful to Christian Krekeler, Thomas Vettorel, Ulf Schiller, Martin Müller, Karen Johnsten, Simon Poblete, Ann Falk and Annette Brunsen for fun conversations making general life at the institute more enjoyable. I also thank all other members of the AK Kremer for insightful questions and comments in seminars and a very enjoyable time at the institute.

I thank the other IRTG members for stimulating scientific discussions and for making both my stays in Korea very pleasent. I especially want to thank Claudine Gross for conversations on computer simulations and for translating countless Korean signs and menus. I also acknowledge Johannes Klos for introducing me to life in Korea during my first stay and Norma Wrobel for providing me with the experimental data on Indolocarbazole.

Thanks also go to my Korean colleagues who had a major part in making my time in Korea a very enjoyable one. I especially thank Sangtaik Noh and Cheol Jeong for helping me with all the organizational issues I had to face in Korea. Nayool Shin and Yeon Sook Chung for scientific cooperation, Yerang Kang for tourguiding and help with countless little things in everyday life, Kyung-Hwan Yoon for excellent organization of all group events and Dukho Kim, Insun Park and Kyong Oh Kim as well as the rest of the group for very nice conversations and cooperation. Finally, I want to thank my friends and family for supporting me throughout my PhD time, especially Carmen Macholdt and Martin Meling for being there when I needed somebody to talk to. I also thank Jorge Cham for making me aware that people around the world seem to face the same difficulties during their PhD.

I want morebooks!

Buy your books fast and straightforward online - at one of world's fastest growing online book stores! Environmentally sound due to Print-on-Demand technologies.

Buy your books online at
www.morebooks.shop

Kaufen Sie Ihre Bücher schnell und unkompliziert online – auf einer der am schnellsten wachsenden Buchhandelsplattformen weltweit! Dank Print-On-Demand umwelt- und ressourcenschonend produziert.

Bücher schneller online kaufen
www.morebooks.shop

KS OmniScriptum Publishing
Brivibas gatve 197
LV-1039 Riga, Latvia
Telefax: +371 686 204 55

info@omniscriptum.com
www.omniscriptum.com

Printed by Books on Demand GmbH, Norderstedt / Germany